Foundations of Principles of Physics

An Introductory Guide

James A. Broberg

Savannah State University

Cover image © Shutterstock, Inc.

Kendall Hunt
publishing company

www.kendallhunt.com
Send all inquiries to:
4050 Westmark Drive
Dubuque, IA 52004-1840

Copyright © 2013 by Kendall Hunt Publishing Company

ISBN 978-1-4652-2595-5

All rights reserved. No part of this publication may be reproduced, stored in a retrieval system, or transmitted, in any form or by any means, electronic, mechanical, photocopying, recording, or otherwise, without the prior written permission of the copyright owner.

Printed in the United States of America
10 9 8 7 6 5 4 3 2 1

Contents

Chapter 1	**Introduction**	1
	What Is Physics?	1
	Scientific Reasoning	2
	Units and Observables	6
	Fundamental Concepts	8
	Problem-Solving Strategies for Checking Your Answers	12
	Homework for Chapter 1	13
Chapter 2	**Review of Algebra**	19
	Single-Variable Problems	20
	Problems with Two Variables	24
	Simplifying Expressions	26
	Basics of Trigonometry	27
	Homework for Chapter 2	29
Chapter 3	**One-Dimensional Kinematics**	43
	Homework for Chapter 3	53
Chapter 4	**Two-Dimensional Kinematics**	71
	Homework for Chapter 4	75
Chapter 5	**Forces and Newton's Laws**	87
	Homework for Chapter 5	99
Chapter 6	**Work and Energy**	125
	Homework for Chapter 6	135
Chapter 7	**Conservation of Linear Momentum and Collisions**	155
	Homework for Chapter 7	165
Chapter 8	**Rotational Kinematics and Dynamics**	183
	Rotational Motion	183
	The Rotational Analogues for the Kinematics Equations	185
	Energy in Rotational Systems	187
	Force in Rotational Systems	191
	Homework for Chapter 8	197

Chapter 9 Coulomb's Law — 213
Vector Arithmetic — 213
Electrostatics: Coulomb's Law — 214
Finding the Electric Field of a Dipole — 221
Electric Potential Energy — 226
Electric Field Lines — 227
Homework for Chapter 9 — 229

Chapter 10 Gauss's Law — 271
Flux and Gauss's Law — 271
Conductors — 273
Capacitance — 274
Homework for Chapter 10 — 277

Chapter 11 Current and Electric Circuits — 283
Current and Drift Velocity — 283
Resistivity, Resistance, and Ohm's Law — 287
Kirchhoff's Rules — 291
Capacitance — 297
Electrical Power — 301
Homework for Chapter 11 — 303

Chapter 12 Magnetostatics — 321
Ampère's Law — 322
Faraday's Law — 323
Lorenz Force Law — 325
Homework for Chapter 12 — 333

Introduction

What Is Physics?

Loosely defined, physics can be understood as being the quantitative study of the fundamental measurable properties of objects in the material world. As such, it distinguishes itself by its *quantitative* rather than descriptive nature from both the humanities and from any premodern inquiries into the nature of reality such as philosophy and mythological explanations. It distinguishes itself from religion and theology by its concern solely with the physical world, and it distinguishes itself from chemistry and biology by its concern with the fundamental nature of matter as such rather than with the specific natures, properties, and behavior of the particular molecule or organism. It distinguishes itself from engineering by its theoretical rather than its practical focus, and from the opposite extreme in mathematics by its empirical base in the actual world rather than the logical relations of pure reason.

This course will approach teaching physics with two main strategies. The first strategy we can call physics for understanding. This strategy is not always employed—many physics problems are expressed by writing algebraic equations in which the variables stand for either measurable quantities or other physical properties of an object or system. Students are given set equations (sometimes also called laws or theorems) and are told to find the value of some physical property by substituting the values that are already known into the equation and solving it algebraically for the missing one. While this is a very good exercise in algebra, it does not give the student a greater understanding of physics without the student knowing why the theorem or law is true and what it means. The first strategy of this textbook will be developing and proving—"deriving"—these laws so that they can be used to solve physics problems, while always stepping back from the mathematics in order to ask ourselves what the equations mean, what the concepts mean, and how they relate to each other. In some of the advanced problems in the course, the student may be asked to work through some simple derivations. Much of the research being done today in physics consists of deriving new laws and equations so that new problems can be solved.

The second strategy will be teaching how to use these given laws and equations in order to solve problems. Solving a computational physics problem sometimes looks intimidating, but it is something a computer could do: One merely needs to follow an algorithm, usually learned unconsciously from years of habit and experience, for determining which equation or physical law reflects the physical situation at hand, determining what the problem is asking for (which variable in what equation to solve for), substituting the correcting known values in the equation and then solving for it. Instead of waiting for you to take 20 years of physics classes, this textbook will teach the rules and methods needed to solve a problem. Physics is difficult because the rules are long—the process for adding two vectors together can be broken down into a dozen or more steps, for example, and determining which method and which equation to use for which problem leaves us with a very long, complicated flow-chart for solving problems. Students often come into classes with difficulty following algorithms—it is more mechanical and less creative than our minds are used to thinking, and the majority of errors are made by a student simply copying down an equation incorrectly between steps 2 and 3 of a problem. Some of the homework problems associated with this chapter are exercises in learning how to follow algorithms and follow instructions. Doing so requires both the attentiveness to read the problem correctly and the patience to focus only on the instruction at hand without jumping ahead to try to solve the problem while skipping the steps in-between. Some students may be surprised to find how difficult they find this act of mental discipline. Everyday thinking tends to employ a "top-down" approach that begins by thinking about what the answer should be and then fleshing out the details later—for example, a student writing a paper for an English or history class will start by writing an outline, and then filling in the text after the ideas have been organized. In a physics problem, one must determine first the physical quantity to be solved for, then determine which law can be used and solve for the desired variable, and then treat every remaining unknown variable as a new problem. The student can make progress only by working through the details, and on challenging problems, the full process needed for solving them through is often not seen until the end.

The top-down organizational approach used in disciplines such as history or political science is a *mental model* that students use to solve the problems in their discipline. In a physics course, students will have to learn entirely different mental models. So our second strategy for this textbook will be to teach the correct mental models. Learning the correct mental models will in part involve learning how to recognize the correct algorithm for solving problems. It will also, however, involve understanding and being able to identify the correspondence between the physical situation posed in a problem and the physical law used to solve the problem. Above all, it involves trying to expedite the training of the student's mind to develop an automatic and reflexive mental association between the physics problem and the correct physical law used to solve it.

In order to start developing the correct mental models and physical intuition to solve a problem, the logic of scientific reasoning needs to be presented so we can understand the underlying method used to solve problems, and also so we can understand how the process of physics problem solving fits into the wider scheme of physics research and our understanding of the universe.

Scientific Reasoning

Scientific reasoning employs two basic forms of argumentation, which were developed in the medieval era and were given the Latin names *modus ponens* ("mode of positing or placing") and *modus tollens* ("mode of taking"). They are called *propositional syllogisms* because the basic units or terms of the argument (syllogism) are propositions—they could be sentences, mathematical equations, or relations and identities. Because the terms are propositions rather than things, they are much more useful for scientific reasoning than are the older *categorical syllogisms* which employed things as its terms (a common example would be "Socrates is a man, all men are mortal, therefore Socrates is mortal," which has a major term "mortal," minor term "Socrates," and middle term "man").

The *modus ponens* argument takes the following form, where p and q represent propositions or complete statements:

$$\text{If } p, \text{ then } q$$
$$p$$
$$\text{Therefore, } q$$

This is usually used when a scientist makes assumptions or simplifications or approximations about a physical situation; these assumptions are called *p*. He then solves the physics problem assuming these approximations ("if *p*, then *q*"), and concludes that the answer he gets is true whenever *p* is true.

P is usually an assumption or approximation, which is true in most situations we would be measuring, or which is close enough to being true that our calculations using the assumption will give us the same answer as we would get if we were to solve the problem the hard way, without using the approximation. For this reason, physics consists of multiple theories and systems, some of which do not give correct answers in extreme conditions, conditions where *p* is not true. Newtonian physics (the subject of the first few chapters of this textbook) is based on a handful of assumptions such as the definition of inertial mass, Euclidean geometry, high temperature, slow speeds, and large sizes. Quantum mechanics and the theory of relativity do not hold to these assumptions, and therefore give different results at small scales and high speeds. But in the everyday world, the difference in the calculations that special relativity gives and in the calculations that Newtonian mechanics gives is too small to measure.

Why do we make assumptions and approximations? The approximations are usually easier to solve; in many cases, the approximations make the problem *possible* to solve. When a difficult problem is faced, such as a description of magnetism or the distribution of electrons in a metal, scientists begin with making simple models of what it looks like, calculating the answers they are looking for and comparing the answers to the experimental results, and gradually making the model more and more nuanced and complex until it matches all of the data.

In an introductory physics class we will not be making new discoveries or developing new models. However, what we will be doing is solving problems assuming certain models, and that is one step of the process of scientific discovery—it is the step "if *p*, then *q*." In the homework sets, we will tie in our problems to actual scientific research (with some made-up scientific research) by giving you models and asking you to calculate the answers to the problems using these models.

The second form of scientific reasoning is the *modus tollens*. This syllogism (logical argument) takes the following form:

If *p*, then *q*
Not *q*
Therefore, not *p*.

This is just as useful as the *modus tollens* in scientific research, and once again the problem solving we will be doing in this textbook will all be examples of the step "If *p*, then *q*." This form of reasoning is used in scientific research to rule out possible models and scenarios. It is also used when trying to measure the properties of certain particles by narrowing down the range of possibilities that need to be measured. Throughout history it has been used in fundamental physics to clarify the definitions of basic physical properties. For example, a *modus tollens* argument can be used to prove that Galileo's definition of acceleration is self-contradictory.

As we will learn later, acceleration in this class is defined as the total change in an object's velocity divided by the change in time. Galileo, working in a pre-Newtonian time period where terminology was still used from Aristotelian physics, would have used the word *motion* to mean "change," so he would have understood our definition of acceleration as the following claim: In any motion [change in velocity] from rest, the speed is proportional to time. Borrowing the later language of the logician Immanuel Kant, this is a synthetic a priori claim—a definition based on more fundamental definitions and which therefore can be either proved or disproved. (Today the most general definition we use for acceleration involves calculus, so we don't employ the Aristotelian notion of "motion" or ideas about proportionality at all. Calculus was invented by Newton, however, roughly a century after Galileo.)

The problem is that Galileo did not say that in any motion from rest that the speed would be proportional to time; he said, incorrectly, that the speed would be proportional to *distance*. Let's use some of our scientific reasoning to show why this cannot be the case.

4 Foundations & Principles of Physics

The (final) speed is still defined as the distance traveled divided by the time it took to travel, and this fact is why Galileo's definition is wrong. Writing down the definition of speed as an algebraic relation,

$$s = \frac{d}{t} \tag{1-1}$$

Let's say d is the distance traveled by an object with uniform speed. Now let's consider Galileo's case of an object undergoing motion from rest—an accelerating body. Because it will not be going as quickly as the object with uniform speed until it reaches that speed at the end, it will go a shorter distance—some distance $\tilde{d}$ when it is at a speed $\tilde{s}$ smaller than s.

Galileo claimed that the speed is proportional to distance, so we write that mathematically as

$$\frac{\tilde{s}}{s} = \frac{\tilde{d}}{d}$$

Substituting the expression $s = \frac{d}{t}$ for s in the above ratio,

$$\frac{\tilde{s}}{\frac{d}{t}} = \frac{\tilde{d}}{d}$$

Solving for $\tilde{s}$,

$$\tilde{s} * t = \tilde{d}$$

$$\tilde{s} = \frac{\tilde{d}}{t}$$

Remember, $\tilde{s}$ and $\tilde{d}$ are speeds and distances of an object that experienced motion *from rest*. But at the beginning of the problem, we stated that $s = \frac{d}{t}$ is a relation that holds true for an object with constant speed. These relations are algebraic where we can plug in any speed or distance for s and d, and we can just as easily take the value for $\tilde{s}$ and plug it in for s and the value for $\tilde{d}$ and plug it in for d, and because $\tilde{s} = \frac{\tilde{d}}{t}$ is true $s = \frac{d}{t}$ will be true using those values—meaning that an object starting from rest and accelerating will go the same distance in the same amount of time as an object starting at the final speed and maintaining uniform velocity. We just argued "if p, then q." But this conclusion is obviously not true: Not q, therefore, not p. Despite Galileo's assumption to the contrary, we have used *modus tollens* to prove that in any motion from rest, the object's (final) speed cannot be proportional to the distance traveled.

Modus tollens is still used today to test hypotheses. Scientists will pose hypotheses and then test them experimentally to see whether their conclusions are true or not; this is the "scientific method" frequently taught in high-school textbooks. Scientific *reasoning* can "test" a hypothesis using purely theoretical arguments as well, by showing the implications of the hypothesis to be absurd (*reductio ad absurdam* is the technical name for this argument) and therefore arguing (using *modus tollens*) that the hypothesis must be incorrect.

For example, 23% of the mass/energy of the universe consists of a mysterious substance called "dark matter," whose gravitational effect can be measured but which does not seem to interact with other matter in any other way. It is hoped that eventually it will be detected in the laboratory, but in order to do so, we must figure out or guess at certain properties, such as the *scattering cross section* (roughly understood, a measurement of the probability of an interaction taking place based on the "target area" that the dark matter particle has to hit). Some of the research the author of this textbook participated in helped rule out possible values for the scattering cross section and then—using *modus tollens*—showing the consequences to be absurd. For example, given a certain range of relationships between the mass of the particle and its cross section, the dark matter particles would lose energy and slow down by hitting protons (hydrogen nuclei) in the upper atmosphere of Jupiter, slowing down enough to be gravitationally captured in the planet's center. I made a second, very probable assumption that these particles are their own antiparticle. These two assumptions together are the premise p. I then calculated the heat that would be caused when these particles annihilate each other ("if p, then q"), and found it to be several times the actual internal heat of Jupiter—"not q." Therefore, dark matter cannot scatter off of protons in the particular range of scattering cross sections that I was looking at.

Above we made the claim that Galileo's definition of acceleration was "synthetic a priori." This classification of propositions is one that employs two distinctions introduced by the logician and physicist Immanuel Kant, writing about half a century after Newton. To roughly paraphrase the thought of an abstruse German philosopher in simple language, Kant divided all possible propositions into those that are simply logical relations developed by the mind independent of empirical experience ("a priori" propositions or propositions true "from before" our observation of the world) and those that are learned from experience ("a posteriori" propositions or propositions true "from after" our observation of the world). He made a second distinction classifying propositions according to whether they were the logical elucidation of meaning already implicit in the definitions given ("analytic" propositions) or whether they were truly informative propositions that give us new information ("synthetic" propositions).

Kant was writing in the wake of a devastating skeptical critique of science and philosophy given by the Scottish empiricist David Hume, who had asserted that all propositions were either what he called "matters of fact" (synthetic a posteriori propositions, or, in less technical language, conclusions drawn directly from observations without inserting any a priori assumptions or definitions into the statement) or "relations of the intellect" (analytic a priori propositions, or pure logical relations, which do not tell us anything new). Such a classification was devastating to science as well as to philosophy, since it meant that scientific claims could not tell us anything new about the world. —Under this scheme, scientific knowledge cannot transcend what we already know from our definitions (used in analytic a priori propositions) or from what we actually tangibly and physically *observe* (in synthetic a posteriori propositions). Science, however, seeks not just to repeat data but to find patterns and make predictions, something impossible to do when all propositions are analytic a priori or synthetic a posteriori. Earlier natural philosophers had tried to assert that science explains the world and gives us true statements about causality—questions of causality were, after all, the whole *raison d'etre* motivating scientific speculation. People want to know *why* the world is the way it is, and this question *why* gave birth to natural philosophy out of which science would blossom. Hume's skepticism claimed that *causality* is not something we can actually directly observe—it is something we *infer* from seeing events occur in succession—and therefore statements about causality are neither synthetic a posteriori or analytic a priori, but rather quite logically meaningless. And indeed, even throughout the nineteenth-century logicians and philosophers of science such as John Stuart Mill would attempt to sidestep the problem of causality and knowledge by evading the syllogism altogether and substituting inductive logic for deductive logic.

Kant was writing in reaction to Hume, whose skepticism, he says, awoke him from his "dogmatic slumber," and tackled Hume's dilemma by the horns by making the claim—which has never ceased to be controversial—that mathematical and physical truths are of a third nature, namely *synthetic a priori*. Physics, of course, depends on observations of the real world in order to motivate the definitions of concepts; we would not have any reason to discuss mass, acceleration, or energy without *seeing* physical realities corresponding to our definitions in the real world. But all of our mathematical equations in this book will be a priori; we will not include a single equation or statement which we are told to accept on faith based on measurement or experiment, as must be done in less fundamental and more complicated branches of science (models in advanced regions of statistical physics, biophysics, astrophysics, and chemistry and biology, for example). We will start with simple definitions and use mathematics to elucidate and combine these definitions to gain as much information as possible from the least number of concepts. It is still empirically grounded because the definitions are of concepts we experience. The mathematical consequences of these definitions are completely a priori, however, and one *could* elucidate mathematical laws based on *any definitions we want* provided they are self-consistent (doing so is not useful for us, however). The only empirical data that will go into our equations will be the values of the physical constants used (the universal gravitational constant, the gravitational acceleration near the earth's surface, the permittivity of free space, etc.), but these are kept as variables in our equations, and the physical *relations* and equations that use them do not depend on observation at all.

Our physical laws will be *synthetic* a priori propositions because they actually do tell us new information about the world. A law such as the law of conservation of mechanical energy, or the law of conservation of linear momentum, is really and truly an informative statement about the world, but it is still simply the logical and mathematical transfigurations of a surprisingly restricted set of basic definitions, and one cannot solve certain physical problems without it. From a physicist's point of view, Kant's insistence that mathematical and physical truths are indeed synthetic a priori propositions is and must be completely true. Those who disagree would try to argue that physical laws are in some way a posteriori instead of a priori, since other models can be constructed using other definitions and assumptions, and it is the *adequation* of the model to reality that judges its merit (hearkening back to the old Aristotelian and medieval definition of truth as the "adequation of the mind to reality"). As this author (who may be biased from his training in mathematics as well as in physics) understands Kant, the development of the model itself given a choice of definitions and assumptions is strictly a priori, not a posteriori, and there is no controversy over the fact that once the definitions and assumptions and "boundary conditions" (restrictions on the mathematics given the physical scenario) are given, the development of these equations follows without any further recourse to laboratory data.

Units and Observables

In making the step, "if p, then q," we take p (usually some equation derived by making an approximation) and solve it for the answer we are looking for. The equations and formulae used in physics have the form of algebraic equations, but unlike the algebraic variables dealt with in mathematical courses, the variables here are physically measurable "observables," measured quantities with some degree of imprecision. Imprecision does not mean inaccuracy; imprecision refers to the level of detail one can obtain from a measurement. A meterstick that measures the length of your wooden block as 5.5 cm is not nearly as *precise* as a Vernier caliper, which measures it to be 5.526 cm, but both are equally *accurate*. The range of uncertainty in every measurement is called *absolute error* and is reported along with the value. The fraction of the absolute error divided by the value reported is the *percentage error*. For example, the measurement found with the meterstick is read 5.5 ± 0.1 cm. We cannot include more digits—called *significant figures*—than we have the precision to distinguish between. It is incorrect to report a number like 7.825 ± 0.1 m/s^2; the last two figures in 7.825 are simply meaningless if the actual value could be anywhere from 7.7 to 7.9 m/s^2, as the range of uncertainty implies. One must instead report the number as 7.8 ± 0.1 m/s^2; both of those numbers are informative and have meaning, and none of the further values do.

For clarity's sake, if the last figure you are allowed to report is in the unit's position (right before the decimal point), place a decimal point *without* any zero afterward to indicate the number of significant figures. For example, 100 has one significant figure, 100. has three, and 100.3 has four significant figures.

The absolute and percentage error ranges are found by taking multiple measurements of the same quantity. In the laboratory, one should always take multiple measurements—usually six to eight—in order to determine the precision of the measurements. All of the measurements should be averaged, and the average value should be reported as the answer. To find the error, subtract the smallest reading in each set of measurements from the largest reading and divide by the average. The value found is called the **fractional error**; multiplying this by 100 gives the percentage error. Multiplying the percentage error by the reported average gives the absolute error, and the absolute error is what needs to be reported with every (averaged) measurement.

When performing calculations with measurements, there are two rules one must follow.

1. When adding and subtracting figures: The number of decimal places in the answer should equal the smallest number of decimal places in the observable that went into the calculation. For example, 5.783 m + 9.54 m = 15.3 m. Do not report 15.323 m.
2. When multiplying and dividing figures: For each observable, calculate what percentage of the value the uncertainty is, then add all of the percentages and use the sum as the percentage uncertainty for the final result. For example, if the length has a 2% error, the width a 4% error, and the height a 5% error, the volume will have an 11% error.

The absolute error should be reported with only one significant figure, unless the digit is 1, in which case two significant figures are used.

When working out a problem, it is best to solve it algebraically and plug in the numerical value of the observable only at the end of the problem, since this shows the physical meaning of the problem more clearly and reduces the error (imprecision) in your answer.

In order to be able to write all of our values with the correct number of digits, all of the observables that we measure and calculate should be expressed in *scientific notation*. A number expressed in scientific notation has one digit followed by a decimal point, then multiplied by 10 to the power of whatever is needed. For example, the number 2,340,000 is written in scientific notation as $2.34*10^6$. This notation helps avoid confusion because 2,340,000 has only three (not seven) significant figures—the zeros at the end are *not* significant, nor are the zeros *before* the first digit in a fraction smaller than 1 (e.g., 0.000076 has only two significant figures, and is written $7.6 * 10^{-5}$).

Every observable is measured in units, and the units themselves must be treated as algebraic variables. In order to ensure consistency in your calculations, it is best to always use the *Systeme Internationale* (SI) system of units. The SI system is also called the "mks" or metric system because its basic units are meters, kilograms, and seconds.

In the metric system, changes of scale are easily made by employing "derivative units," which are multiples of 10 of the basic units and are named by a set list of prefixes attached to the basic units. For example, one-hundredth of a meter is a centimeter, one-hundredth of a gram is called a centigram, and one-hundredth of a second would theoretically be called a "centisecond" (not a term in common usage). One thousand meters is a kilometer, and 1000 grams is a kilogram. The most commonly used prefixes together with the multiple of 10 from the basic unit that it corresponds to in scientific notation is given in the table below.

Prefix	Pronunciation	Multiple of Basic Unit
f	femto-	$1*10^{-15}$
n	nano-	$1*10^{-9}$
μ	micro-	$1*10^{-6}$
m	milli-	$1*10^{-3}$
c	centi-	$1*10^{-2}$
d	deci-	$1*10^{-1}$
k	kilo-	$1*10^{3}$
M	mega-	$1*10^{6}$
G	giga-	$1*10^{9}$
T	tera-	$1*10^{12}$

Fundamental Concepts

The most basic concept in physics is position or *displacement*. A revolution in scientific thinking came about shortly after Galileo when the philosopher/scientist René Descartes, lying in bed at night staring at the ceiling, noticed that the entire space of the ceiling could be mapped with two coordinates measured by orthogonal axes. Position is measured by coordinates, and, as we'll later see, is a *vector*; it is treated as a line going from the origin of the coordinate system out to the place where the position is measured. Because it is a vector, the difference in position in two-dimensional space can be calculated numerically, and this difference is called *displacement*. A change or difference in any value is defined by taking the final value and subtracting the initial one, and the change is indicated by the Greek letter "Δ" or delta. Displacement is then defined as

$$\Delta x \equiv x_f - x_i \qquad (1\text{-}2)$$

where the subscripts f and i always indicate final and initial values.

Displacement can either be measured in the laboratory (or, for our purposes, given in a problem), or solved for in a problem. Physics problems that ask for displacement will use questions such as "how far?" "how high?" "where will it be?" or similar phrases. It is important to learn to recognize that when a problem asks a question like this, we should be solving a law of physics for displacement.

The equations are occasionally written with x_f and x_i instead of Δx, in part because of the way the equations are derived using calculus. But the coordinate system is arbitrary—it does not matter whether you say that the ground is where an object is at a height of 0 meters or whether the top of the hill is declared to be 0 meters, *so long as all subsequent measurements of position are measured relative to the same starting point or origin*. In other words, only Δx has actual physical meaning; you can change the location of the origin of the coordinate system in any problem and you will get the same answer, because all problems use the coordinates only by taking the difference between them.

Not only does it not matter which location you take to be the origin of the coordinate system, it does not even matter if the coordinate system is *moving*. There is no "thing" or substance called space; displacement is simply a measurement of the distance between two objects, and when those two objects are moving relative to each other, it is completely arbitrary at which one we measure the origin of the coordinate system. (One might object and say "we measure it on the ground," but the ground is a thing, too. And while in practice we *do* usually measure the origin of the coordinate system somewhere on the ground, we forget that the ground is moving quite rapidly—over a thousand miles an hour if you are near the equator—because the earth is spinning on its axis.) If one were driving in a car or vehicle of any other sort that was *perfectly* smooth—no bumps in the road, no vibration from the car engine, no turbulence—then we could walk around in it and perform delicate balancing tasks, and everything would be the same as if we were still. If we look outside, we can see that we are moving, but it is only our past experience or bias that tells us that "the earth is a 'fixed reference point,' therefore, if the two are moving relative to each other it is the train or car that is doing the motion," and the reason for this is that the train is much smaller and therefore slows down from the force of friction at a much, much faster rate than the earth slows its rotation down (not to mention that for every train slowing the earth down in one direction, another train going the opposite direction cancels out the effect), and consequently it is the train, not the earth, that needs to keep putting fuel in its engine. As far as physics is concerned, however, either the train or the earth is equally valid as a "reference frame," and either could be considered to be the one in motion, since they are both in motion *relative to each other*. Motion only has meaning *relative to* two objects or between one object and an arbitrary reference frame, and the physical equations we will develop will hold true in *all* classical reference frames in which the reference frame itself is not accelerating with respect to the object in question—what we have come to call an *inertial reference frame*. This important fact is called the *Galilean principle of relativity*, because it was Galileo who proposed it (Einstein's principles of relativity is a much stronger, and completely different, claim).

Once a reference frame or coordinate origin is chosen, the displacement is measured as the distance between the origin and the point in question. Common units of displacement found in everyday life are inches, feet, yards, centimeters, miles, and light years. (The definition of "everyday life" is a bit flexible for science-fiction buffs and theoretical astrophysicists.) In physics, we want to ensure consistency by only using the *Systeme Internationale* or mks unit of displacement, which is the meter.

The first attempt to provide a precise, rigorous standard for the meter was in 1668 when an English scientist by the name of John Wilkins defined it as the length of a pendulum whose half-period is one second. The period of a pendulum depends on its length, and Wilkins's genius lay in his use of physical properties to relate our unit of length to a unit of time based on the natural cycles of the earth's rotation. However, in 1791, the French Academy of Sciences redefined the meter as being one ten-millionth of the meridian of a quadrant, which is one ten-millionth of the distance from the earth's equator to the North Pole at sea level.

It is difficult to measure the distance from the equator to the North Pole in those everyday life situations where one needs an absolutely precise definition of the meter, so, in 1889, a platinum-alloy bar was engraved with two notches by the International Bureau of Weights and Measures (*Bureau international des poids et mesures*) and placed in a vault kept at 0° C in Paris. The distance between the two notches became the official, conventional definition of a meter. That definition lasted until the 1960 *Conference Generale des Poids et Mesures* (General Conference of Weights and Measures) attempted to return to a natural (rather than conventional) definition of a meter by defining it as 1,650,763.73 wavelengths of the orange-red emission line in the electromagnetic spectrum of Kr 86 in a vacuum. In 1983 that same conference redefined the meter as "the length of the path traveled by light in a vacuum during a time interval of 1/299792458 seconds," which is the most current official definition of a meter.

Having discussed pretty much all there is to say about distance in and of itself without getting into nonstandard unit systems and exotic coordinate systems (your instructor will probably save the polar coordinates for a later class, and if you beg hard enough he or she might give you hyperbolic and elliptical ones, too), it's time to talk about the next basic concept in physics: change.

Change is always change *of* something. In ancient times change was called "motion," and was divided into many subdivisions which do not concern us here because of their qualitative nature. Modern physics does not care about change in "substance" or in qualitative properties, but in measurable physical properties. Change in position would also have been considered to have been "motion" by Aristotle—as it is known in common language today—but hidden within this terminology is the implicit assumption that there is something absolute or fixed about our coordinate system (an assumption that runs contrary to our physics that obeys Galilean relativity). If it is only the distance between two points or displacement that has physical meaning and not the position of those points themselves, then we should not be using "motion" as a fundamental concept in physics. Someone moving at a steady rate or *inertial frame of reference* is, from his perspective, motionless. Except for buffering from the wind and air resistance and vibration from poor suspension, people cannot *feel* any motion in a moving car, train, or airplane except when they are turning or speeding up or slowing down. From their own point of view, they are at rest.

Having introduced the concept of change, we need to ask what it is that is changing. So far the only other concept we've mentioned is position, and we've seen what the change in position is—displacement, or Δx. Δx could be the length of an object, but more often in physics we are interested in the change of position for one object changing locations. It takes time to do that, and so the third basic concept to be introduced is time.

The SI unit of time is the second, which, of course, comes down to us from the old Babylonian sexagesimal system, whereby all measurements came in packages of sixty or other multiples of six. We have the day divided into twenty-four hours, the hours divided into 60 minutes, and the minutes divided into 60 seconds. Modern science could not fathom being so irrational as to define a unit by something as arbitrary as division into 60 parts, so the new definition of the second given to us in 1997 by the International Bureau of Weights and Measures is "the duration of 9,192,631,770 periods of the radiation corresponding to the transition between the two hyperfine levels of the ground state of the cesium 133 atom at rest at temperature K = 0."

The next fundamental concept in physics is mass. Newtonian physics in a very real sense saw its birth with Newton's realization, confirmed experimentally thousands of times, that inertial mass and gravitational mass are proportional and therefore can be defined as equivalent. These concepts will be discussed in more depth later on. For now, we can think of mass as a measurement of the amount of "stuff" in something. Mass is *not* the same thing as weight—weight is a gravitational force, and changes depending on one's altitude or on which planet you are on. The mass of an object is the same everywhere in the universe, in all places, at all temperatures, and at all speeds. Special relativity nuances this statement slightly—Einstein discovered that it was "mass-energy" rather than mass which is the fundamental constant, and the Nobel laureate Richard Feynman attempted to redefine momentum as the most basic physical concept resulting in a "relativistic mass" depending on the particle's velocity. Feynman's suggestion has been almost universally rejected, however. Physicists have grown up with our mothers teaching us from the cradle that mass is the measurement of the amount of stuff in something, and so Feynman or no Feynman it *must* remain constant, at all costs.

The SI unit of mass is the kilogram. Students with a strong background in chemistry must be careful here, because the gram is typically the basic unit for chemistry. In physics, all of the physical constants are defined using kilograms as the basic unit whenever mks units are employed, and very few publications in physics still use the "cgs" (centimeters-grams-seconds) unit system any more. The kilogram was originally defined as the mass of 1 liter of distilled water at 0° C, a definition that dates to the sunny and bloody days of Revolutionary France; today, the official definition of the kilogram is given as the mass of a certain platinum cube held in a vault in Paris.

Meters, kilograms, and seconds will be the basic units used for most of this book. There are other SI units that will be introduced later in the text when they become relevant. As a teaser, we can mention some of the units to be seen in the topics in physics later in the book. Electricity and magnetism, which form the bulk of a second semester of introductory physics, use the ampere as the fundamental unit of current. Optics uses the candela as the fundamental unit for intensity.

For the rest of our discussion on mechanics, we will only be using units derived from meters, kilograms, and seconds, called *composite concepts*. Units can be multiplied and divided by each other just like ordinary algebraic variables, although quantities with unlike units cannot be added or subtracted (doing so has no physical meaning—what does a meter plus two seconds mean anyway?). As such, we can derive new units for new physical quantities by multiplying and dividing units by each other.

When units are encountered in non-SI systems, it is convenient to convert them to SI units so that they can be used consistently with other SI units. To do this we employ *unit conversion factors*. Since units can be multiplied and divided (though not added or subtracted) from each other just like any algebraic variables, we can divide two identical units with each other to get 1, called a "unit conversion factor." Any quantity with units A is converted to any quantity with units B by multiplying by the unit conversion factor $\frac{bB}{aA}$, where a and b are the coefficients that make the unit conversion factor equal to 1. The unit A will cancel out, leaving the unit B.

For example, to convert 17 inches to meters, we use the following facts: there are 2.54 cm in one inch, and 100 cm in one meter. Arranging unit conversion factors so that inches cancel out leaving centimeters and then centimeters cancel out leaving meters,

$$17 \text{ in} * \frac{2.54 \text{ cm}}{1 \text{ in}} * \frac{1 \text{ m}}{100 \text{ cm}} = 0.432 \text{ m}.$$

We mentioned previously that displacement is change in position, usually the change in one particle's position as it moves through time. The *average velocity* of a particle is defined loosely as the rate at which the particle is moving, or more rigorously as

$$v \equiv \frac{\Delta x}{t} \qquad (1\text{-}3)$$

where t is the time in-between the final and initial measurements, and could also be written Δt (although the Δ is conventionally not written for time in order to make equations a lot simpler to write).

In order to find the units of velocity, we simply take the units of the two quantities divided by each other and then divide the units; all of the units on the left side of the equation must match all of the units on the right side. Displacement has units of meters, and time has units of seconds, so velocity has units of meters/second (m/s). 1 m/s is roughly 2 miles/hr.

Let's be careful and clarify a potential confusion here. Displacement is the final position minus the initial position; wherever the object was in-between *does not affect* the displacement. A person driving around in a circle to end up where she started will have a displacement 0. Displacement, in other words, is not th e same thing as path length—in fact, it *will not* be equal to the path length except when the motion is in one direction in a straight line. Likewise, the average velocity for circular motion that returns to its starting point will be 0. This sounds paradoxical, but velocity is not speed—speed is the *magnitude* of velocity. Velocity is a vector, meaning that it has both a magnitude and a direction, and the car's velocity while it is going forward will be canceled out by its velocity when it is going backwards. In one dimension, forward motion is described with positive velocity, and motion heading back toward the origin is described with negative velocity. Speed, on the other hand, is always positive, and is defined as being the path length (not the displacement) divided by the time.

We are often interested in the velocity of an object at a given moment, rather than the average velocity over a spread of time. We do this by writing the definition of average velocity explicitly as a function of a change in time,

$$v \equiv \frac{\Delta x}{\Delta t}$$

Vand then taking the limit $\Delta t \to 0$, or letting the time considered become arbitrarily small, so that we are looking at an instantaneous snapshot of the particle's velocity. When we take the limit, the deltas become derivatives, and the definition of instantaneous velocity is

$$v \equiv \frac{dx}{dt} \qquad (1\text{-}4)$$

Since velocity is the rate of change in position, we can define another basic quantity by finding the rate of change in velocity. This is *acceleration*, a concept not to be confused with *velocity*. Velocity tells us how fast one is going; acceleration tells us how the velocity is changing. Average acceleration is defined in a manner similar to average velocity,

$$a \equiv \frac{\Delta v}{\Delta t} \qquad (1\text{-}5)$$

and instantaneous acceleration is also found by taking the limit as $\Delta t \to 0$:

$$a \equiv \frac{dv}{dt} = \frac{d^2 x}{dt^2} \qquad (1\text{-}6)$$

The units on the left must match the units on the right, and if we are dividing velocity by time, the units will be m/s divided by s, or m/s^2.

Finally, area and volume have units derived only from the basic unit of length, m. The area of a rectangle is the base times its height, where both the base and height have units of meters, so in order for the units of area to match the other side of the formula, area must have units of m^2. Likewise, volume has units m^3.

Problem-Solving Strategies for Checking Your Answers

Even though we have not introduced any physics yet, from the discussion we've already presented, we can develop strategies that will be used in all of our problems to check our work and to make sure that our answers are reasonable.

1. Dimensional analysis. As mentioned above, the units on both sides of the equation must match. The two sides of the equation are literally the same thing, expressed differently; their units must be the same. One meter cannot be the same as three seconds. Therefore, one can always check the accuracy of one's answer by seeing if the units on both sides of the equation match.
2. Order of magnitude estimates. All of the observables that we measure and calculate should be expressed in *scientific notation* in order to clarify the number of significant figures. A quick and dirty way of checking the accuracy of an answer is to ignore the value of the digit and add the exponents of 10 multiplied together and subtract the exponents of 10 that are divided; what we will get is the exponent of the final answer. Rounding up numbers like $9 * 10^9$ to $1 * 10^{10}$ before doing so will increase the accuracy of the estimate. If your calculated answer is something 1000 times higher, you probably did something wrong.
3. Always make sure your answer makes physical sense. Often the mathematics will give you answers that you do not need and are not relevant to the problem. For example, a quadratic equation will always give you two solutions. If one of them is a solution for negative time or negative mass or negative distance and the question is clearly asking for a positive value, then do not use the second solution; simply ignore it.

HOMEWORK FOR CHAPTER 1

Name _____

A. The following exercises are meant to give the student practice following instructions and carrying through algorithms. These skills will be used for the rest of the book. These problems are meant to be simple. The student does not need to know what any of the equations *mean*; we simply want the student to learn how to follow instructions.

1. Write the following equation down with "+5" inserted on both sides of the equals sign after all the other terms:

$$5x - 5 = 7$$

2. Rewrite the following equation writing down everything on the left side, writing an equals sign, and then copying down everything on the right side:

$$v^2 - v_0^2 = 2a\Delta x$$

3. Rewrite the following equation replacing "v_0" with "17" and replacing "a" with "-9.8". Do not change anything else:

$$v = v_0 + at$$

4. Follow these instructions:
   a. Solve the following equation for ω by rewriting it multiple times following the instructions given below:

   $$\frac{7\partial\varepsilon}{\partial\overleftrightarrow{v}} + \sqrt{17\breve{\alpha} + e^{-i\omega t}} = -\nabla\hat{\varphi}$$

   i. Insert $-\dfrac{7\partial\varepsilon}{\partial\overleftrightarrow{v}}$ immediately to the left of the equals sign and immediately to the right of the expression on the right side of the equals sign.

   ii. Rewrite the equation you have in the first step by moving the $\dfrac{7\partial\varepsilon}{\partial\overleftrightarrow{v}}$ immediately next to $-\dfrac{7\partial\varepsilon}{\partial\overleftrightarrow{v}}$ on the left side of the equation, without adding anything not found in step I or changing anything else.

14   Foundations & Principles of Physics

iii. Rewrite the equation you have in step II replacing $\dfrac{7\partial\varepsilon}{\partial \overleftrightarrow{v}} - \dfrac{7\partial\varepsilon}{\partial \overleftrightarrow{v}}$ with 0 in the expression. Do not change anything else.

iv. Rewrite the equation you have in step III leaving out "+0." Do not change anything else.

v. Enclose everything on the right-hand side of the equals side with parentheses. Do not change anything else.

vi. Rewrite the equation you have in step V without the square root sign over the terms on the left-hand side of the equation, and add a superscript ^{2} to the right of the terms on the right-hand side of the equation. Do not change anything else.

vii. Rewrite the equation you have in step VI inserting $-17\breve{\alpha}$ immediately after the already existing $17\breve{\alpha}$, and insert $-17\breve{\alpha}$ to the right of everything on the right-hand side of the equals sign. Do not change anything else.

viii. Rewrite the equation you have in step VII replacing $17\breve{\alpha} - 17\breve{\alpha}$ on the left side of the equals sign with 0. Do not replace anything else.

ix. Rewrite the equation you have in step VIII leaving out the "0 +." Do not change anything else.

x. Write down ln($) = ln(&), but instead of $ write down everything on the left-hand side of the equals side in the equation in step IX, and instead of & write down everything on the right-hand side of the equals side in the equation in step IX. Do not change anything else.

xi. Take the equation you have in step X and replace everything on the left-hand side of the equals sign with just the group of characters on the left-hand side with superscripts. Do not change anything else.

xii. Take the equation you have in step XI and divide both sides of the equals sign by $-it$. Do not change anything else.

xiii. Switch the order of ω and t in the expression on the left-hand side of part xii.

xiv. On the left-hand side of the equals side in the equation you obtained in step XII cross out the expression $\frac{-it}{-it}$ found there. Do not change anything else.

xv. Rewrite the equation obtained in step XIII omitting the crossed-out expression. Do not change anything else.

xvi. Rewrite the equation obtained in step XIV replacing the $-i$ in the denominator of the right-hand side of the equals sign with an i in the numerator of the right-hand side of the equals sign. This expression is our final result. This is the sort of algebraic maneuvering you will be expected to be able to do on your own to solve the homework problems in this book.

B. The following simple questions are on units and unit analysis.
1. The period of a simple pendulum, defined as the time necessary for one complete oscillation, is measured in time units and is given by

$$T = 2\pi\sqrt{\frac{l}{g}}$$

where l is the length of the pendulum. Use algebra to solve for the units of g.

2. Force has SI units of kg*m/s². Knowing this,
   a. What are the units of the spring constant k in Hooke's law, $F = -kx$ where F is a force and x is the displacement?

b. What are the units of the frequency of a simple harmonic oscillator, given by the equation $f = \dfrac{1}{2\pi}\sqrt{\dfrac{k}{m}}$ where m is the mass of the spring and k is the spring constant?

c. What are the units of the potential energy of a spring stretched to a displacement x, given by the formula $U = \dfrac{1}{2}kx^2$?

d. Coulomb's law tells us that for some charges q_1 and q_2 with units "C" (coulombs) and a distance r between them, the electric force between the two particles is $F = k_e \dfrac{q_1 q_2}{r^2}$. Find the units of k_e.

e. k_e is sometimes written as $k_e = \dfrac{1}{4\pi\varepsilon_0}$ where ε_0 is a constant called the "permittivity of free space." Find the units of the permittivity of free space.

f. Newton's law of universal gravitation is given by

$$F = G\dfrac{Mm}{r^2}$$

where M and m are the masses of two objects (usually a planet or star and a satellite or object on the surface) and r is the distance between the center of masses of the two objects. What units does the proportionality constant G have?

3. State which of the following equations are dimensionally consistent.
   a. $v = v_0 + at^2$

   b. $x_f = x_0 + v_0 + \frac{1}{2}at^2$

   c. $v_0 t = 7\dfrac{Fm}{k^2}$ where k is the spring constant, m is a mass, and F is the force.

   d. $\Delta x = \frac{1}{2}(v + v_0)t$

4. Linear momentum is defined as mass times velocity, $p = mv$.
   a. Give the SI units for linear momentum.

   b. Just by rearranging the quantities so that the units match, write an expression giving force as a function of momentum and time. This is actually how equations are often derived—by matching the units and then measuring a "proportionality constant" making the numbers equal. In this case, the proportionality constant is 1.

   c. In problem 2, part (c) we derived the units for the potential energy of a spring. All energy has the same units. By matching the units, find an expression for energy as a function of momentum and mass. The proportionality constant is $1/2$.

5. A cardboard box has sides 5.21 m, 7.3 m, and 40 m. Find the surface area and volume of the box retaining the correct number of significant figures.

6. The permittivity of free space has been measured to be $8.854187817620 \times 10^{-12}$, in units derived in an exercise above. Express this number with (a) three significant figures, (b) eight significant figures, (c) nine significant figures, and (d) twelve significant figures.

7. A calculation in a physics problem gives you an answer $7389247029304829834 \pm 2\%$. Express this number in scientific notation using the correct number of significant figures.

8. How many significant figures are in the following numbers?
   a. 40 m

   b. 0.004 km

   c. 49.5 cm

d. 27 m²

e. 40. furlongs

9. The SI unit system is also called the mks system because it uses meters, kilograms, and seconds. We could use other unit systems. The cgs (centimeters, grams, seconds) system has already been mentioned. The FFF (furlongs, firkins, and fortnights) system is used less often, usually only by computer scientists and professors with a penchant for schadenfreude. To convert from the SI unit system to the FFF system, we use the conversion factors 1 furlong = 201.168 m, 1 firkin = 40.8233133 kg, and 1 fortnight is two weeks.

   a. Express the gravitational acceleration, -9.80 m/s², in furlongs per fortnight squared, using the correct number of significant figures.

   b. If weight is mass times gravitational acceleration, what is the weight of a 90-kg table in the FFF system?

Review of Algebra 2

In physics, there are three basic mathematical tools at our disposal to solve physics problems: algebra, trigonometry, and calculus. This textbook will present a noncalculus-based approach to physics, an approach that works when certain symmetries in the physics problems reduces the calculus to trivial calculations performable with algebra. Algebra and trigonometry are the heart and soul of physics, however, and it is essential to have an absolutely masterful grasp of both disciplines.

This chapter will be brief and will skip the theoretical principles underlying algebra—the law of associativity, commutativity, etc.—traditionally presented in a pre-algebra course, and will assume that the student has some familiarity (which may have become buried in old memories!) of the subject equivalent to a high-school sequence of Algebra I and Algebra II. Instead of repeating that, we shall take an algorithmic approach to presenting a refresher on the topic. We will assume that the student is comfortable with the concept of a variable—a sign, usually represented by a letter, denoting an unknown quantity which must be isolated or "solved for" from all the other quantities in an equation through manipulating symbols following the rules of algebra—and with the concept of an equation. We will spend the rest of this chapter giving concise lists of rules to follow for solving different types of algebraic problems. The chapter will end with two more essential but basic topics, simplifying expressions and basic trigonometry definitions and methods. Physics will usually not involve the use of complicated or clever trigonometric identities, only the basics.

There are three types of equations we will encounter in physics, those with one variable to be solved for, those with two variables to be solved for, and those with three variables to be solved for. (Problems with four or more unknown variables are very tedious to solve, and are easiest solved using techniques from a discipline called *linear algebra*, which is beyond the scope of this book. Such problems, including those encountered by the author of this book and his colleagues in the course of research, can involve a million or more variables, and are solved by computers.)

There are two methods for solving two-variable problems, called "elimination" and "substitution," and problems with three or more variables are extensions of these methods. Single-variable problems are solved according to different methods depending on which type of equation is given, and we will again subdivide these problems into two types of equations: linear and quadratic. The algorithm for solving quadratic equations depends on whether or not they have a linear term, and also on whether or not they have a constant term. Taking all of these divisions of the types of algebra problems, we can present an outline for the different types of problems encountered, and present simple algorithms for each entry on the outline.

A. Single-variable problems
   a. Linear equations
   b. Quadratic equations
      i. Quadratic equations with a constant term but no linear term
      ii. Quadratic equations with a linear term but no constant term
      iii. Quadratic equations with both linear and constant terms
         1. Solution by factoring
         2. Solution by the quadratic formula (completing the square)
B. Two-variable problems
   a. Method of substitution
   b. Method of elimination
C. Three-variable problems

Let's follow the rest of the chapter according to this outline.

Single-Variable Problems

For any algebraic problem, the number of unknown variables we are solving for must be equal to the number of independent equations used to solve the problem. In other words, if, for example, a physics problem asks one to find both the mass and final velocity of an object in a collision, one must be able to write down two separate equations from the information given in the problem in order to find the two unknown quantities. If the problem is only asking for the acceleration of an object in a certain situation, then we only need one equation—an equation that contains acceleration (which we would solve for) as well as other physical observables that we are already given.

So single-variable problems all reduce in the end to solving single algebraic equations, and, as we noted above, these equations can be *linear* or *quadratic*.

A *linear* equation is one in which the unknown variable never appears in powers higher than 1; in other words, if x is the variable we are solving for, we are never going to see an x^2 or x^3 or x^{50} in the equation. Granting that we won't see terms like that, graphing the dependent variable (the one we are looking for) against the independent variable (anything else in the equation) for any equation produces a straight line, so these equations are called *linear*. (Conceptually, a line is actually called such because it is produced by a linear equation—the equation is a more basic and fundamental concept than the graph. But when first learning the subject it's easier to understand lines than equations.)

A typical example of a linear equation is the following:

$$3x + 5 = 7$$

Here clearly the variable we are solving for is x — we know what everything else is. But it frequently happens in physics that we end up with equations where all or most of the symbols in the formula are symbolic letters rather than numerals, and we know the values of all of them except for one. So, the following equation is also a linear equation:

$$v = v_0 + at$$

This is an equation from physics relating the final velocity of an object ("v") to its initial velocity ("v_0"), the acceleration that causes the velocity to change ("a"), and the time frame between the initial and final states ("t"). A common problem given to students is for them to be asked, given a ball thrown upward at some velocity ("v_0"), how long will it take for the ball to reach its peak. "v_0" is given in the problem, and physical facts common to all free-projectile motion tell us what "v" and "a" are. This is, for the sake of this problem, a linear equation whose variable is "t."

In order to solve an equation, we want to manipulate the symbols in such a manner as to get the variable by itself on the left side of the equation, with everything left over on the right side being the answer we are looking for. So, in the examples above, we want the left sides of the equations to read "$x =$ " and "$a =$ ". There are two rules for manipulating equations. You can add, subtract, multiply, or divide *anything you want* (except divide by 0) to the equation provided that

1. Whatever you do is done to the *whole side* of the equation – it must be done to *everything* on that side of the equals sign; and
2. Whatever you do must be done to *both sides* of the equation.

In addition, you can modify *just one side* of the equation by either multiplying it by 1 (which doesn't change anything!) or by adding zero (which also doesn't change anything!), but 1 and 0 can be expressed in as clever a manner as you could want, since any nonzero quantity divided by itself is 1 and any quantity subtracted from itself is 0.

In other words, if you have the equation $3x + 5 = 7$ and you want to divide by 3, you have to divide every term by 3—the $3x$ gets divided by 3, the 5 gets divided by 3, and the 7 gets divided by 3.

Granting that given those two rules everything is permitted, not everything that *can* be done necessarily *should* be done. Not everything is useful. One might fumble around for hours—most likely going around in circles—without isolating the variable, all without doing anything mathematically incorrect. So let's give a useful algorithm for solving these equations as quickly as possible.

We want to isolate the x in the equation $3x + 5 = 7$, which means getting rid of both the 3, which multiplies the x, and the 5, which is added to it. "Getting rid" of a numeral means performing the inverse function to that by which the numeral is present, so if x is multiplied by 3, we want to divide by 3; and if x is added to 5, we want to subtract 5. It actually doesn't matter which one you do first. But in order to avoid ugly factors and make life simpler, as a general rule, *always get rid of the terms added to the variable first*. In other words, subtract 5 from both sides of the equation.

Step 1. Subtract all of the terms added to the variable from both sides of the equation.

$$3x + 5 = 7$$
$$3x + 5 - 5 = 7 - 5$$
Since $5 - 5 = 0$,
$$3x = 7 - 5$$

Now we notice that this can be written in a simpler fashion, since $7 - 5$ is an easy arithmetic problem. Rewriting $7 - 5$ as 2 is called *collecting like terms*; in other words, collecting like terms is adding anything together that you can compute arithmetically.

Step 2. Collect like terms.

$$3x = 7 - 5$$
$$3x = 2$$

Now would be a good time to divide both sides by 3. Doing so will cancel out the 3, which multiplies by the variable, leaving x alone on the left side. All of the terms multiplying by the variable we are looking for are called the variable's *coefficients*; it is simplest to divide by the coefficients last so that we could have already collected like terms and have fewer fractions to deal with.

Step 3. Divide both sides by the coefficients of the variable.

$$3x = 2$$
$$\frac{3}{3}x = \frac{2}{3}$$

Since $\frac{3}{3} = 1$, $x = \frac{2}{3}$.

This algorithm can be used to solve *any* linear equation with one variable, and will yield one and only one solution.

The second type of equation is called a *quadratic* equation, and is defined by the presence of a term in which the variable is squared. For example, we might be given the equation

$$3x^2 + 4x = 7$$

In this example, $3x^2$ is called the "quadratic term" because of the x^2 in it, and $4x$ is called the "linear term." If there is no linear term, this can be solved the same way as we solve a linear equation—just treat x^2 as being the variable, solve for x^2, and then take the square root of both sides of the equation. Unfortunately it isn't this simple when a linear term is present. We haven't been told how to solve linear equations that have $\sqrt{x}$ in them, and that's what you'd get by treating x^2 as the variable in such an equation. (Actually, the best way to solve linear equations with $\sqrt{x}$ in them is to square both sides to make them quadratic—solving quadratic equations is just easier.)

If the constant (the term that doesn't have any "x" in it) is 0, then we can find two solutions to the quadratic equation relatively simply. (All quadratic equations have two solutions.) One of the solutions is going to be 0, since every nonzero term is going to have x in it, and 0 times anything is 0. For example, you can easily check and see that 0 works as a solution for x in the equation:

$$3x^2 + 4x = 0$$

On the other hand, we can divide any equation or number by anything which is *not* 0. So granting that 0 is one option for a solution, let's suppose $x \neq 0$ and divide both sides of the equation by x. That gives us

$$3x^2 + 4x = 0$$
$$3x + 4 = 0$$

which is an easy linear equation to solve. To refresh our memories on how to solve linear equations, let's work it out quickly below:

$$3x + 4 = 0$$
$$3x + 4 - 4 = 0 - 4$$
$$3x = -4$$
$$\frac{3}{3}x = \frac{-4}{3}$$
$$x = \frac{-4}{3}$$

In most cases, however, neither the linear term nor the constant will be 0, and we'll be stuck with equations such as $3x^2 + 4x = 7$. In order to solve equations like this, we must put them in *quadratic form*, or in the arrangement $ax^2 + bx + c = 0$, where a, b, and c are the coefficients of the variable x.

We can easily put the equation $3x^2 + 4x = 7$ into quadratic form by subtracting 7 from both sides:

$$3x^2 + 4x = 7$$
$$3x^2 + 4x - 7 = 7 - 7$$
$$3x^2 + 4x - 7 = 0$$

Once an equation has been put into quadratic form, there are two methods we can use, one easier but not always available, and the other longer but always reliable.

The easier method is called factoring. It can't be done very intuitively with the equation we have given, so let's take a simpler quadratic equation:

$$x^2 + 5x + 6 = 0$$

In order to factor this (or any other) equation, we must find two numbers (call them j and k) such that $j + k = b$ where b is the coefficient of the linear term, and $j * k = c$ where c is the constant. The coefficient of the quadratic term (a) must be 1, which can be obtained easily by dividing every term in the equation by a.

In this case, we want two numbers that add to 5 and multiply together to get 6. The numbers 3 and 2 meet this criterion. To solve an equation by factoring, we declare that $(x + j)(x + k) = 0$, which will always be true if $a = 1, j + k = b$, and $j * k = c$. (Multiply those two terms together using the FOIL method, and add like terms to verify this.)

Now if two terms multiplied by each other equals 0, then one or the other of the two terms must be 0. You can't have two nonzero terms multiplied together to get 0 for the reason that division by 0 is impossible. So either $x + j = 0$ or $x + k = 0$. And these are both linear equations, which yield the solutions $x = -j$ and $x = -k$ respectively. (Again, a quadratic equation will *always* yield two solutions, something we must be careful about, because in physics only one of them usually makes any sense.)

It's not always easy to factor an equation though (although it's always *possible*—just make $-j$ and $-k$ be the solutions, which you'd have to actually find using the method presented next). For equations such as $3x^2 + 4x - 7 = 0$, a more general method has to be used. Early mathematicians developed a method for solving equations like this called *completing the square*, which is more or less a method to rewrite these equations as perfect squares (sets of terms that can be factored into a linear expression multiplied by itself) with a constant left over on the other side, taking the square root of both sides, and then solving the linear equation. That method can be done with the general formula $ax^2 + bx + c = 0$, which when solved for x gives us the famous *quadratic formula*, a formula that tells us x for *any* quadratic equation:

$$x = \frac{-b \pm \sqrt{b^2 - 4ac}}{2a} \quad (2\text{-}1)$$

The "$\pm$" means plus or minus and is an indicator that there are two solutions: for one of them, the symbol has to be a "$+$" and for the other, the symbol has to be a "$-$".

Let's do an example of solving a quadratic equation using the quadratic formula.

$$3x^2 + 4x - 7 = 0$$

$a = 3, b = 4$, and $c = -7$

So $x = \dfrac{-4 \pm \sqrt{4^2 - 4(3)(-7)}}{2(3)}$

$$x = \frac{-4 \pm \sqrt{16 + 84}}{6}$$

$$x = \frac{-4 \pm \sqrt{100}}{6}$$

$$x = \frac{-4 \pm 10}{6}$$

$$x = \frac{6}{6} \text{ or } x = \frac{-14}{6}$$

$$x = 1 \text{ or } x = \frac{-7}{3}$$

Check and verify that either of these works as a solution to the equation.

We won't usually be seeing equations with x^3 or x^4 in them, and there is no good simple way to solve them. The steps provided for solving equations of one variable already will suffice for all the one-variable problems you will encounter in this book.

Problems with Two Variables

The key to solving problems with two variables it to be able to reduce them to single-variable equations, and then solve them using the methods already discussed. There are two methods that can be used: substitution and elimination.

Substitution

For any problem with two variables, we'll need two equations to solve them. (If you graph each of the two equations, using one variable as the dependent variable and the other one as the independent variable, then the solution to the system of equations is where the two lines intersect.) The trick to *substitution* is to solve one of the equations (call it equation 2) for one variable *in terms of the other one*, substitute that expression into equation 1 to find what the other variable is, and then plug that back into equation 2 to find the value of the first variable. An example will serve well to illustrate.

$$y + 6 = 4x$$

$$\frac{7x}{2} + y = 9$$

This is a problem with two equations and two unknowns: x and y. Let's solve the top equation for y and plug it back in to the bottom equation, which will then be an equation that is only in terms of x:

$$y + 6 = 4x$$
$$y + 6 - 6 = 4x - 6$$
$$y = 4x - 6$$

Now substitute $4x - 6$ for y in the bottom equation:

$$\frac{7x}{2} + y = 9$$

$$\frac{7x}{2} + 4x - 6 = 9$$

We want to isolate x, so we add 6 to both sides and then collect like terms:

$$\frac{7x}{2} + 4x - 6 = 9$$

$$\frac{7x}{2} + 4x - 6 + 6 = 9 + 6$$

$$\frac{7x}{2} + 4x = 9 + 6$$

Rewriting $4x$ as $\frac{8x}{2}$,

$$\frac{7x}{2} + \frac{8x}{2} = 9 + 6$$

$$\frac{15x}{2} = 15$$

Now we isolate x by multiplying both sides by the reciprocal of its coefficient (since multiplication by the reciprocal of a fraction is the same as division by that fraction):

$$\frac{2}{15} * \frac{15x}{2} = 15 * \frac{2}{15}$$

The 15s and 2s cancel out, and we are left with

$$x = 2$$

That's part of the answer to the problem, but we're not quite done yet—we know x, and we need to also find y. So now we go back to the equation that was solved for y in terms of x:

$$y = 4x - 6$$

and enter in the value of 2 for x:

$$y = 4x - 6$$
$$y = 4(2) - 6$$
$$y = 2$$

Elimination

The second method, which is usually harder but sometimes desirable, employs adding the two equations together in order to eliminate one of the variables. Since you can add or subtract anything to both sides of the equation as long as you add *the same thing* to both sides, it doesn't matter whether that same thing is written the same way, or whether they are two sides of another equation (an equation gives an equality, so you're still adding the same thing to both sides of an equation by adding another equation to it).

We'd do this if you see obvious ways in which the coefficient of one of the variables in one equation is a multiple of the coefficient of that same variable in the other equation. Let's illustrate with an example.

$$3y + 4x = 7$$
$$5y + 2x = 9$$

Now 4 is obviously a nice easy multiple of 2, since $2 * 2 = 4$. They're both coefficients of x, so x is the variable we shall be eliminating. In order to do this, multiply one of the two equations by some constant so that the coefficients of x in the two equations are *equal and opposite* to each other, and will cancel out; in this case, multiply the second equation by -2.

$$3y + 4x = 7$$
$$-2(5y + 2x = 9)$$

$$3y + 4x = 7$$
$$-10y - 4x = -18$$

These are still the same two equations we started with, because you can do *anything you want* to one side of an equation as long as you do the same to the other side, and the equation remains the same. Now let's add the two equations together:

$$3y - 10y + 4x - 4x = 7 - 18$$

Adding them together gives us only one equation now, but that's okay—it will have only one variable after we've collected like terms, and we can go back to the original equations to solve for the other variable. Collecting like terms,

$$-7y = -11$$

Dividing both sides by -7,

$$\frac{-7}{-7}y = \frac{-11}{-7}$$

$$y = \frac{11}{7}$$

Now we can go back to either of the original equations—it doesn't matter which—and substitute $\frac{11}{7}$ for y and solve for x:

$$3y + 4x = 7$$

$$3\left(\frac{11}{7}\right) + 4x = 7$$

$$\frac{33}{7} + 4x = 7$$

$$\frac{33}{7} - \frac{33}{7} + 4x = 7 - \frac{33}{7}$$

$$4x = 7 - \frac{33}{7}$$

$$4x = \frac{49}{7} - \frac{33}{7}$$

$$4x = \frac{16}{7}$$

$$\frac{4}{4}x = \frac{16}{7} * \frac{1}{4}$$

$$x = \frac{4}{7}$$

The following steps summarize the algorithm for solving systems of two equations using elimination.

Step 1: Look for a variable in one equation that is a simple multiple of the same variable in the other equation.

Step 2: Multiply the second equation by the *negative* of that multiple.

Step 3: Add both sides of the two equations to each other.

Step 4: Solve for the variable that is left over.

Step 5: Plug the value for this variable back into either one of the original equations and solve for the other variable.

Simplifying Expressions

In addition to solving equations, physical fluency requires a core competency in a handful of other algebraic tricks and facts, including the ability to simplify expressions. In general, there is no "one right way" to simplify an expression—it is more of an art than a science. However, these four steps provide a simple and reliable algorithm to follow:

Step 1: Multiply everything out.

Step 2: Collect like terms, including putting everything over a common denominator if necessary.

Step 3: Factor the numerator and denominator of the single fraction you have.

Step 4: Cancel out whatever possible.

Basics of Trigonometry

We'll cover the trigonometry in more depth when it's needed; for the first few chapters, all we need to know are the basic definitions and their use. Trigonometry is built on the fact that similar triangles have sides with the same proportions; therefore, a ratio of the sides of one right triangle will be the same as the ratio of the sides of *any* right triangle with the same angles. These ratios are the basic trigonometric functions, called sine, cosine, and tangent:

$$\text{Sin } \theta = \frac{\text{opposite leg}}{\text{hypotenuse}} \tag{2-2}$$

$$\text{Cos } \theta = \frac{\text{adjacent leg}}{\text{hypotenuse}} \tag{2-3}$$

$$\text{Tan } \theta = \frac{\text{opposite leg}}{\text{adjacent leg}} \tag{2-4}$$

θ is a general symbol usually used to denote an angle; although defined (at least originally) by the ratios of triangle legs, these are actually functions of angles, and can be defined (using Euler's theorem in complex analysis) completely irrespective of triangles at all.

These functions are used in physics because many physical quantities—displacement, velocity, acceleration, force, momentum, fields—are vectors. A vector is a quantity with a magnitude and a direction; it is usually represented by an arrow starting at the origin whose length is its magnitude and whose direction is its angle from the *x*-axis, rotating counterclockwise. As we'll see when we begin to solve multidimensional kinematics problems, in order to perform computations with vectors, we need to break the vector up into its *components*. If the vector is an arrow starting at the origin, then the components are its projection along the *x* and *y* axes, the legs of a right triangle of which the magnitude is the hypotenuse.

Treating the magnitude of the vector as the hypotenuse of its components allows us to develop a simple geometrical method for going back and forth between the expression of a vector as a magnitude and direction and its decomposition into components. The Pythagorean theorem, which is a special case of the law of cosines, tells us that for any right triangle the length of the hypotenuse *c* is related to the lengths of the sides *a* and *b* by the relation

$$a^2 + b^2 = c^2 \tag{2-5}$$

So for any vector **A**, the magnitude A is found from the components as

$$A = \sqrt{A_x^2 + A_y^2} \tag{2-6}$$

Other relations are found using the trigonometric functions. Since

$$\text{Sin } \theta = \frac{\text{opposite leg}}{\text{hypotenuse}}$$

for a vector pointing at an angle θ from the *x*-axis,

$$\text{Sin } \theta = \frac{A_y}{A} \tag{2-7}$$

where A is the magnitude of the vector and A_y the y-component (or the projection of the vector along the *y*-axis).

Therefore, solving for A_y, we find that $A_y = A \sin \theta$
Likewise,

$$\text{Cos } \theta = \frac{A_x}{A} \tag{2-8}$$

and $A_x = A \cos \theta$.

When given the components of a vector and asked to find the direction θ, use the *inverse* trigonometric functions. As would be expected from an inverse function, these are defined as

$$\theta = \text{Sin}^{-1} \frac{\text{opposite side}}{\text{hypotenuse}} \tag{2-9}$$

$$\theta = \text{Cos}^{-1} \frac{\text{adjacent side}}{\text{hypotenuse}} \tag{2-10}$$

$$\theta = \text{Tan}^{-1} \frac{\text{opposite side}}{\text{adjacent side}} \tag{2-11}$$

The inverse tangent function is the easiest to use to find the direction of a vector given its components, but once the magnitude is known, any three can be used:

$$\theta = \text{Sin}^{-1} \frac{A_y}{A} \tag{2-12}$$

$$\theta = \text{Cos}^{-1} \frac{A_x}{A} \tag{2-13}$$

$$\theta = \text{Tan}^{-1} \frac{A_y}{A_x} \tag{2-14}$$

Finally, there are three special triangles one should be familiar with. A **45-45-90** right triangle has angles 45°, 45°, and 90°, and its legs are of equal length with a hypotenuse $\sqrt{2}$ times the length of each leg. A **30-60-90** right triangle has angles 30°, 60°, and 90°. The leg opposite the 60° angle is $\frac{\sqrt{3}}{2}$ times the length of the leg opposite the 30° angle, and the hypotenuse is twice the length of the leg opposite the 30° angle. Finally, a 3-4-5 right triangle has sides of proportions 3, 4, and 5, with the angles being irrational numbers.

HOMEWORK FOR CHAPTER 2

Name_____

A.

1. Solve for x: $3x = 6$

2. Solve for b: $5b = 7$

3. Solve for y: $(8 * 10^7)y = 2.7 * 10^9$.

4. Solve for v: $(2.7 * 10^4)av = \sqrt{(7x)}$

5. Solve for x: $4x = 6$

6. Solve for b: $8b = 7$

7. Solve for y: $29y = 37$

8. Solve for v: $(17 + x)v = 25g$

B.

1. Solve for x: $3x + 5 = 6$

2. Solve for b: $5b + 3 = 7$

3. Solve for y: $(8 * 10^7)y = 2.7 * 10^9 + v_0^2$

4. Solve for v: $(2.7 * 10^4)av + 7v_0^2 = \sqrt{(7x)}$

5. Solve for x: $4x + 7 = 9$

6. Solve for b: $12b + 4 = 2$

7. Solve for y: $(7 * 10^9)y - 17\Delta q = 4\pi z^2$

8. Solve for a: $v = v_0 + at$

C.

1. Solve for x: $\dfrac{3}{x} = \dfrac{5}{y}$

2. Solve for y: $\dfrac{3}{x} = \dfrac{5}{y}$

3. Solve for ΔV: $C = \dfrac{Q_{tot}}{\Delta V}$

4. Solve for a: $v^2 = v_0^2 + 2a\Delta x$

5. Solve for t: $x_f = x_i + at^2$

6. Given the following three equations, use substitution to express σ in terms of l, C, ΔV, and w:

$$\sigma = \frac{Q_{tot}}{A}$$

$$A = lw$$

$$C = \frac{Q_{tot}}{\Delta V}$$

D.

1. A vector **A** has magnitude 25 at an angle 25°. Find the x and y components of **A**.

2. A vector **B** has components $17x + 18y$. Find the magnitude and direction of **B**.

3. A vector **C** has magnitude $\sqrt{2}$ at an angle 45°. Find the x and y components of **C**.

4. A vector **D** has components $5x + 5\sqrt{3}y$. Find the direction and magnitude of **D**.

One-Dimensional Kinematics 3

In Chapter 1 we defined a number of basic concepts and terms, describing physical *observables*, that our equations are supposed to relate to each other. In this chapter, we will explore the concepts of displacement, time, velocity, and acceleration to develop four useful equations for solving problems involving motion. The study of motion is called *kinematics* from the Greek word *kinema*, meaning motion, and the equations we will be deriving are developed strictly from the definitions we've given of displacement, velocity, acceleration, and time. (So for those of you who are philosophically inclined and are up to speed on your Kant, or who still remember Chapter 1, these equations are synthetic a priori propositions—even though the *application* of any of these equations involves empirical measurement to find the values of the observables, the *definitions* of which are given a priori. Enough philosophy for this chapter!)

Let's revisit the definitions of the basic concepts – of our fundamental observables – given in chapter 1. The most basic concept we dealt with was displacement, defined as *change in position* irrespective of path length. Change in position is written mathematically as

$$\Delta x \equiv x_f - x_i \quad (3\text{-}1)$$

and since the initial and final points must be measured using some coordinate system (for this book, a Cartesian coordinate system will usually work best), the initial and final positions can be defined as vectors beginning at the origin and ending at the point. So displacement is the difference between two vectors, and is itself a vector—*path length* or the total length traveled, by contrast, is a scalar.

The second basic concept was *average velocity*, also a vector and given by

$$v \equiv \frac{\Delta x}{t} \quad (3\text{-}2)$$

(Δx is a vector, and when one divides a vector by a scalar one gets another vector, the length of which is "scaled" by whatever you multiply it by, hence the name "scalar." So v is a vector. The magnitude of this vector is its speed—and be careful, because average speed and average velocity could have completely different values.)

And thirdly, we divide by time again to get another vector, average acceleration:

$$a \equiv \frac{\Delta v}{\Delta t} = \frac{\Delta v}{t} \quad (3\text{-}3)$$

Note that we use the symbols "t" and "Δt" interchangeably. Like displacement, time only has physical meaning when we're talking about a *change* in time. There isn't any "absolute time" on some cosmic clock somewhere. We could start looking at a problem whenever we want, and the laws of physics aren't going to change. For simplicity, the change in time, which could be thought of as being the time elapsed, is usually just called "t."

Let's not worry about the instantaneous velocity and acceleration for a moment, since they involve calculus, and just think about the average values.

Let's start with the definition of acceleration.

$$a \equiv \frac{\Delta v}{t}$$

Recalling that $\Delta v = v_f - v_0$, this definition can be rewritten

$$a \equiv \frac{v_f - v_0}{t} \tag{3-4}$$

Remember that physics is interested in *observables*. These values are important because we measure them. In particular, in any given problem, the final value is the value you are measuring now—the initial velocity v_0 is some value that was measured earlier, or given to you. So if a problem is asking you what the velocity is given information about the state of the particle beforehand, we are looking at the final velocity in a process, and we might as well just relabel the final velocity v_f as being the measured velocity, v. This is a convention almost always used. So we can again rewrite the definition of acceleration using v instead of v_f, as

$$a \equiv \frac{v - v_0}{t}$$

Acceleration isn't what we're usually looking for. We're more typically interested in more directly measurable quantities, like velocity or time. Equation 3-4 is a linear equation that we can solve for (final) velocity relatively simply enough. First, multiply both sides by t.

$$at = v - v_0$$

And then to find v, add v_0 to both sides of the equation:

$$v = v_0 + at \tag{3-5}$$

That was almost trivial (synthetic a priori propositions always are—until they get *long* to work out, and length is a matter of opinion), and yet we've developed an equation useful for solving problems.

Let's take an example. Say you are throwing a ball up in the air with an initial velocity of 8 m/s. How long will it take to reach the peak of its trajectory?

In order to use equation 3-5 to solve these, you need to know two facts about free projectile motion. The first is that the acceleration of an object due to gravity near the surface of the earth is *always* -9.8 m/s^2, regardless of the object's mass, size, shape, or material. (Other forces may act on it such as air resistance, which is why a feather floats to the ground while a cannonball drops, but that's not acceleration due to gravity alone.) This isn't an intuitive fact: Aristotle believed that heavier objects dropped faster than lighter ones, and it took Galileo (in an apocryphal story) to prove that all objects experience the same gravitational acceleration by dropping cannonballs of different masses off of the Leaning Tower of Pisa.

The second fact you need to know is that the velocity (or, in two dimensions, the *y*-component of the velocity) of *any* free projectile at its peak is 0. Remember that velocity is a vector — it's always positive when it's going up, and negative when it's going down. At the peak, it's in-between positive and negative, and so it must be 0.

So to use equation 3-5 to solve this problem, we plug in 0 for v, 8 m/s for v_0, -9.8 m/s^2 for a, and solve for t. It's always best to solve a physics problem algebraically and then plug in the numbers at the end so you can see the actual physical relationships all the way through the problem rather than just the numbers. So let's start by solving equation 3-5 for t.

$$\mathbf{v} = \mathbf{v_0} + \mathbf{a}t$$

$$\mathbf{a}t = \mathbf{v} - \mathbf{v_0}$$

$$t = \frac{v - v_0}{a} \tag{3-6}$$

Strictly speaking I did something which a mathematician would start griping about. All of the quantities in **bold print** are vectors, and all the quantities in *italics* are scalars—that's a typical typographical convention. For the last step, they all became scalars. Why? There's no definition for what it means to divide by a vector. In order to get to the last step, we had to replace the vectors with their magnitudes, which are scalars, and we'll only get the same answer doing this if all of the vectors lie in the same direction. As it turns out, that's okay, because you can only arithmetically add vectors pointing in the same direction, and so in order to add vectors in general (as we'll be doing in the next chapter), you need to break them up into their *components*. Until we encounter the "cross product" in later chapters, motion in the *x* and *y* directions will be independent of each other, and the actual equations we end up using for multidimensional problems are the equations using just the components of the vector equations derived in this chapter. But this is a point one might not expect a beginning physics student to catch, and it is worth noting: this derivation only works because the problem is one-dimensional.

Now we plug in the values we know for the three quantities on the right-hand side of the equation, and calculate *t*.

$$t = \frac{0-8}{-9.8} = \frac{8}{9.8} = 0.816 \text{ sec}$$

So far so good. Since we've derived something useful and informative out of the definition of average acceleration, let's see what we can get by playing around with the definition of average velocity.

$$\mathbf{v}_{ave} \equiv \frac{\Delta \mathbf{x}}{t} \tag{3-7}$$

This is *average* velocity now, not *final* velocity, even though in the definition given above we just used "*v*." For the sake of clarity here, we'll keep the subscript "*ave*" to indicate that we're talking about the average velocity. Let's say we want to find the displacement, Δx. We can isolate it easily by multiplying both sides of the equation by *t* to get

$$\Delta \mathbf{x} = \mathbf{v}_{ave} t \tag{3-8}$$

Now for uniform acceleration, the average velocity is going to be half of the initial velocity plus the final velocity—right in-between the two, in other words. So using

$$\mathbf{v}_{ave} = \frac{1}{2}(\mathbf{v} + \mathbf{v_0}) \tag{3-9}$$

where once more *v* indicates final velocity,

$$\Delta \mathbf{x} = \frac{1}{2}(\mathbf{v} + \mathbf{v_0})t \tag{3-10}$$

which is the second kinematics equation.

Now equation 3-10 has *v* in it, and in equation 3-5 we derived an expression for *v*. The next logical step therefore is to plug the expression for *v* found in equation 3-5 into equation 3-10 to see what we get.

$$\Delta \mathbf{x} = \frac{1}{2}(\mathbf{v} + \mathbf{v_0})t$$

$$\Delta \mathbf{x} = \frac{1}{2}(\mathbf{v_0} + \mathbf{a}t + \mathbf{v_0})t$$

$$\Delta x = \frac{1}{2}(2v_0 + at)t$$

$$\Delta x = \frac{1}{2}(2v_0 t + at^2)$$

$$\Delta x = v_0 t + \frac{1}{2}at^2$$

Now remembering the definition

$$\Delta x \equiv x_f - x_i$$

we see that

$$x_f - x_i = v_0 t + \frac{1}{2}at^2$$

Since initial velocity in this equation is called v_0, it makes sense to rename the initial position x_0. Also, since in equation 3-5 the final velocity was called v; likewise, it makes sense to rename the final position—*the position which will be measured, or the position that a question would ask for*—as simply x. In this case

$$x - x_0 = v_0 t + \frac{1}{2}at^2$$

For reasons that become apparent when seeing how this equation is derived using calculus, it is usually rearranged to be written

$$x = x_0 + v_0 t + \frac{1}{2}at^2 \qquad (3\text{-}11)$$

This is the third kinematics equation.

Finally, let's write an equation which doesn't have time in it. Equation 3-5 can be solved for time (in terms of acceleration and initial and final velocities, both of which are already present in this equation) and plugged in here. Earlier we saw that equation 3-5 solved for time was

$$t = \frac{v - v_0}{a}$$

Substituting this for time in equation 3-11 gives us

$$x = x_0 + v_0 \frac{v - v_0}{a} + \frac{1}{2}a\left(\frac{v - v_0}{a}\right)^2$$

Simplification is our friend. For one thing, let's write the displacement as Δx again:

$$x - x_0 = v_0 \frac{v - v_0}{a} + \frac{1}{2}a\left(\frac{v - v_0}{a}\right)^2$$

$$\Delta x = v_0 \frac{v - v_0}{a} + \frac{1}{2}a\left(\frac{v - v_0}{a}\right)^2$$

Now remember that the expression we substituted for time only works if something is accelerating in the direction of motion. Taking that condition again in order to do something that will make a professional mathematician tear his or her hair out, let's cancel the vector **a** in the numerator with the scalar a in the denominator. Once again, if vectors are going in the same direction, they add the same way as scalars do, so this is okay; from the point of view of mathematical rigor, we've skipped a lot of steps in showing that we can do this (and under what conditions we can do this). But physicists usually do something mathematically dubious and then let mathematicians prove that it works later, later usually meaning on the order of decades to centuries. For our purposes, this cancellation is perfectly fine for a one-dimensional problem.

$$\Delta x = v_0 \frac{v - v_0}{a} + \frac{1}{2}\frac{(v - v_0)^2}{a}$$

We can put everything on the right-hand side of the equation over a common denominator by multiplying the first term by $\frac{2}{2}$.

$$\Delta x = \frac{2}{2}v_0 \frac{v - v_0}{a} + \frac{1}{2}\frac{(v - v_0)^2}{a}$$

Writing them over the common denominator,

$$\Delta x = \frac{2}{2}v_0 \frac{v - v_0}{a} + \frac{1}{2}\frac{(v - v_0)^2}{a}$$

$$\Delta x = \frac{2v_0(v - v_0) + (v - v_0)^2}{2a}$$

Now simplifying by multiplying everything out (and using a convenient sleight of hand to replace the magnitude v_0 for the vector $\mathbf{v_0}$—we can do this for the same reason we misbehaved all the other times, because it's a one-dimensional situation),

$$\Delta x = \frac{2v_0 v - 2v_0^2 + v^2 - 2v_0 v + v_0^2}{2a}$$

Since the author is too self-respecting a physicist to leave the left-hand side of the equation a vector while the right side is a scalar (scalars can't be equal to vectors), we made the displacement a scalar, too. For a one-dimensional problem, we can do that. The expression we have can be simplified by collecting like terms.

$$\Delta x = \frac{v^2 - v_0^2}{2a} \tag{3-12}$$

This equation is often written in a number of different ways and could be fine and useful the way it is. Let's pander to the author's nostalgia for his college days, when it was given as a function of v^2.

$$v^2 - v_0^2 = 2a\Delta x$$
$$v^2 = v_0^2 + 2a\Delta x \tag{3-13}$$

For the mathematically inclined readers of this text, note that all of the terms here are *supposed* to be scalars, mathematical shortcuts aside. All of the products are dot products or "inner products" between vectors, and the dot product between two vectors is a scalar. Specifically, v^2 is $\mathbf{v} * \mathbf{v}$, v_0^2 is $\mathbf{v_0} * \mathbf{v_0}$, and $a\Delta x$ is $\mathbf{a} * \mathbf{\Delta x}$. For the sake of simplicity, however, they're usually written as scalar products as given here.

Let's summarize the four equations we've developed so far, and then give some rules for discerning when to use what in a physics problem.

1. $v = v_0 + at$
2. $\Delta x = \frac{1}{2}(v + v_0)t$
3. $x = x_0 + v_0 t + \frac{1}{2}at^2$
4. $v^2 = v_0^2 + 2a\Delta x$

Often in a physics problem you will be given a choice as to which equation to use. The rule of thumb is to solve for a variable you *don't* know, using an equation that contains information you *do* know. The following three rules are typically a good place to start:

1. If the problem neither gives you nor asks you for any information about the time, use equation 3-13.
2. If the problem neither gives you nor asks you for any information about the final velocity, use equation 3-11.
3. If the problem neither gives you nor asks you for any information about the displacement, use equation 3-5.
4. If the problem neither gives you nor asks you any information about the acceleration, use equation 3-10.

Let's see how these equations are used in a few examples.

Example 3-1. Find the zenith (maximum height) reached by a free projectile with an initial velocity 5 m/s.

A free projectile is one acting only under the influence of gravity. As discussed earlier, all objects near the earth's surface (irrespective of size, mass, or physical/chemical makeup) experience an acceleration due to gravity equal to -9.8 m/s^2, with the negative sign indicating that the acceleration goes downward. (The number 9.8 m/s^2, without the negative sign, is called "g.") The maximum height of a projectile is when it stops going upward (which entails positive velocity) and starts going downward (which means negative velocity); in other words, when the velocity is 0. So, for this problem we know the initial velocity (it is given to us, 5 m/s), the final velocity (0 m/s), the acceleration (-9.8 m/s^2), and we want to find the maximum height reached, or, in other words, the displacement. Time has not been mentioned, so we use equation 3-13 and solve for Δx.

$$v^2 = v_0^2 + 2a\Delta x$$
$$0 = (5)^2 + 2(-9.8)\Delta x$$
$$2(9.8)\Delta x = 5^2$$
$$19.6\Delta x = 25$$
$$\Delta x = 25/19.6 = 1.28 \text{ m}$$

Example 3-2. Find the length of time it takes for the free projectile with an initial velocity 5 m/s to reach its maximum height.

Now we are asked for the time, and we know the displacement, initial velocity, final velocity, and acceleration. (We calculated the displacement in Example 3-1, and all the conditions of the problem were the same, so the answer holds true in this problem as well.) Equation 3-13 is the only one that does not have time in it, so we are free to use any of the other three. Equation 3-5 is the simplest of the other three, but we can show how it can be solved using all three of them.

1. Using equation 3-5,

$$v = v_0 + at$$
$$0 = 5 + (-9.8)t$$
$$9.8t = 5$$
$$t = 5/9.8 = 0.51 \text{ s}$$

2. Using equation 3-10,

$$\Delta x = \tfrac{1}{2}(v + v_0)t$$
$$1.28 = 0.5(0 + 5)t$$
$$1.28 = 0.5 * 5t$$
$$1.28 = 2.5t$$

$$t = 1.28/2.5 = 0.51 \text{ s}$$

3. Using equation 3-11,
$$x = x_0 + v_0 t + \tfrac{1}{2}at^2$$

Remember that the displacement $\Delta x = x - x_0$, since "x" here is the final position. We can label the initial position 0 (since we can choose any coordinate system we want, as long as the *difference* between points remains the same) and then be able to use the displacement as the final position.

$$1.28 = 0 + 5t + 0.5(-9.8)t^2$$

Moving all the terms over to the left side of the equation,

$$0.5(9.8)t^2 - 5t + 1.28 = 0$$
$$4.9t^2 - 5t + 1.28 = 0$$

This is a quadratic equation that can be solved using only the quadratic formula,

$$t = \frac{-b \pm \sqrt{b^2 - 4ac}}{2a}$$

a is the coefficient of the quadratic term, in this case 4.9. *b* is the coefficient of the linear term, in this case -5. C is the coefficient of the constant, in this case 1.28.

$$\text{So } t = \frac{+5 \pm \sqrt{(-5)^2 - 4(4.9)(1.28)}}{2(4.9)} = (5 \pm \sqrt{(25 - 25)}))/9.8 = 5/9.8 = 0.51 \text{ s}$$

Example 3-3. Let's say you throw two balls off a cliff 25 m high, one straight up with a velocity 5 m/s, and another straight down with a velocity -5 m/s. Which will have a higher speed when it lands at the bottom of the cliff?

Time wasn't mentioned, so we know right off the bat that we need to use equation 3-13. We're looking for final velocity (or actually, the magnitude of final velocity, speed, without caring which direction it's going), and we know acceleration (-9.8 m/s^2, as always for free projectiles), initial velocities, and the displacement (final position minus initial—here $0 - 25 = -25$ m for both balls). We can only describe the motion of *one* ball per each equation, so we'll need to use equation 3-13 twice and solve for the final velocities separately, and then compare.

For the first ball,
$$v^2 = v_0^2 + 2a\Delta x$$
$$v^2 = (-5)^2 + 2(-9.8)(-25)$$
$$v^2 = 25 + 490$$
$$v^2 = 515$$
$$v = -22.7 \text{ m/s}$$

(The final velocity is the *negative* square root of 515 because it is going down, and all motion in the negative direction is negative.)

Now for the second ball,
$$v^2 = v_0^2 + 2a\Delta x$$
$$v^2 = (5)^2 + 2(-9.8)(-25)$$
$$v^2 = 25 + 490$$
$$v^2 = 515$$
$$v = -22.7 \text{ m/s}$$

Whoa! The final velocities were the same. That's because equation 3-13 is a function of displacement, which doesn't care about the path length—only the initial and final positions matter. It will take much longer for the second ball to get down to the bottom, but equation 3-13 doesn't say anything about time: it only tells you that when it finally does get there, it will have the same speed as the first one did. Later on, we'll see that this is an example of the principle of *conservation of mechanical energy*.

Example 3-4. Let's say an airplane lands on a runway with an initial speed of 71.5 m/s, and slows to a stop with an acceleration of −4.47 m/s. How much runway space will it need to land?

First comment: acceleration can mean *any* change in velocity, whether it's just a change in direction around a circle, or speeding up (in which case the acceleration is positive), or slowing down (in which case the acceleration is negative). In ordinary language we don't usually use "acceleration" to mean slowing down, but this is one case in which the technical use of a term in physics is different from its use in ordinary language.

Time hasn't been mentioned, so we need to use equation 3-13 again, this time solving for displacement (since we want to know how long the runway needs to be) rather than velocity. The final velocity will be 0, since we need the airplane to come to a rest so the passengers can disembark.

$$v^2 = v_0^2 + 2a\Delta x$$
$$2a\Delta x = v^2 - v_0^2$$
$$\Delta x = (v^2 - v_0^2)/2a$$
$$\Delta x = (0 - 71.5^2)/(2*-4.47)$$
$$\Delta x = -5112.25/-8.94 = 572 \text{ m}$$

Example 3-5. A stone is dropped off a cliff. After 3.00 sec of free fall, what is its displacement?

We're told time and we want to find displacement; the only thing not mentioned was final velocity. Consequently, we must use equation 3-11.

$$x = x_0 + v_0 t + \tfrac{1}{2}at^2$$

"x" is normally used to denote position along the *x*-axis, or the horizontal direction. Here we are dealing with vertical motion, motion along the *y*-axis, so without changing the truth of the equation any we can replace the x with y (since these laws of motion don't depend on whether it's horizontal or vertical motion you're talking about—the physics is the same, since the coordinate axes were defined arbitrarily).

$$y = y_0 + v_0 t + \tfrac{1}{2}at^2$$

We want to know the displacement, or Δy, which equals $y - y_0$ as before, so we rewrite the equation as:

$$y - y_0 = v_0 t + \tfrac{1}{2}at^2$$
$$\Delta y = v_0 t + \tfrac{1}{2}at^2$$

The initial velocity is 0, since it was just sitting in your hand before it was dropped (in other words, it was *dropped*, not thrown). The acceleration is the acceleration due to gravity, $-g$.

$$\Delta y = 0t + 0.5(-9.8)t^2$$
$$\Delta y = 0.5(-9.8)(3^2) = -44.1 \text{ m}$$

So the final position of the stone is 44.1 m below the initial position of the stone.

Example 3-6. A car is moving down a highway at 20 m/s, and accelerating at 2 m/s². How far away from its present position will the car be after 5 sec?

Again, time has been mentioned, and we want to find the displacement, and no mention is made of final velocity, so we have to use equation 3-11. This time, since we're going in the horizontal direction, we'll use x instead of y.

$$x = x_0 + v_0 t + \tfrac{1}{2}at^2$$
$$x - x_0 = v_0 t + \tfrac{1}{2}at^2$$
$$\Delta x = v_0 t + \tfrac{1}{2}at^2$$

This time, $t = 5$, v_0 is given to us as 20, and $a = 2$.

$$\Delta x = 20(5) + (0.5)(2)(5^2) = 100 + 25 = 125 \text{ m}.$$

HOMEWORK FOR CHAPTER 3

Name _____

A. A ball is thrown upward in one dimension with an initial speed of 17 zlugs per megahandles, near the surface of the planet Earth. (It does not matter what a zlug or a megahandle is. In all honesty, I actually just made them up.)

   1. What is v_0? *Hint:* This is not a trick question—it is as trivial as it looks.

   2. What is the ball's velocity at the peak of its trajectory?

   3. What is the ball's final velocity just before it hits the ground?

   4. What is the ball's acceleration in m/s² at the peak of its trajectory?

   5. What is the ball's acceleration in m/s² at the beginning of its trajectory?

   6. What is the ball's acceleration in m/s² at the end of its trajectory?

B. Consider the equation $x_f = x_0 + v_0 t + \frac{1}{2}at^2$.
   1. What are the units of x_f?

   2. What are the units of x_0?

   3. What are the units of v_0?

   4. What are the units of t?

   5. What are the units of a?

C. A question asks how far something went in 20 seconds given an initial velocity of 5 m/s and an initial acceleration of 3 m/s². Which variable in the equation are we supposed to solve for, and what units will it have?

D. A question asks how long it took a projectile to fall from rest from a height of 17 m. Which variable in the equation are we supposed to solve for, and what units will it have?

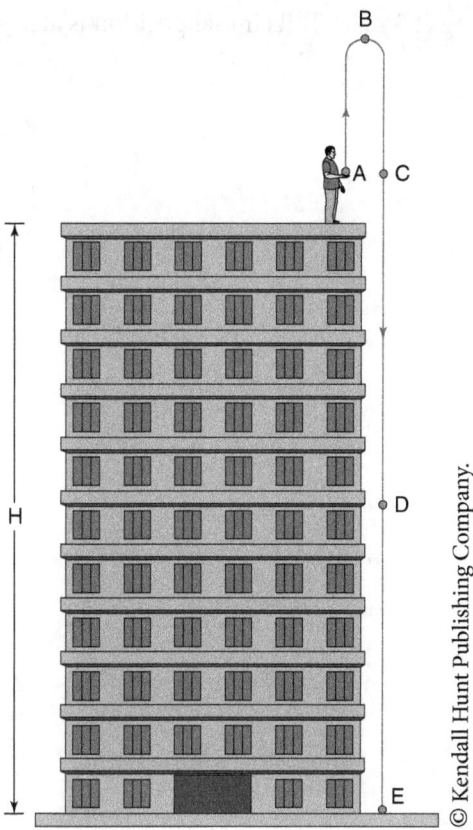

E. A friend standing on the roof of a 9-story building with height H throws a ball upward with an initial speed of 5 m/s.

1. At which location(s) (A, B, C, or D), if any, is the velocity 0?

2. At which location(s) (A, B, C, or D), if any, is the acceleration 0?

3. On your paper, draw the direction of velocity and acceleration vectors at A, B, C, and D. Mark them all clearly.

4. What is the difference in height between the ball at point A and at point B?

5. What is the velocity of the ball at point C? Show your work

6. What is the velocity of the ball at point D? Express your answer in terms of H.

F. A car is moving down a highway at -15 m/s and accelerating at 3 m/s². If its initial position is at $x = 0$ m, where will it be after 3 seconds?

G. A jet plane has a takeoff speed of $v_{to} = 125$ km/h (note the units!) and its engines can power it to accelerate with an average acceleration of 5.6 m/s².
   1. What length of runway will it need to take off safely?

2. Suppose it had an available runway length of only 75 m. What minimum constant acceleration would it need to take off safely?

3. Supposing the runway length and acceleration in question 2, how long will the plane take to take off?

4. Upon arriving at its destination, the plane lands with a speed of 500 km/h and its brakes can cause it to accelerate at -4.5 m/s^2 without severely compromising the comfort of the passengers. How long will the airplane take to come to rest?

5. Suppose the runway is 1,000 yards long. Will the plane be able to land safely?

H. Delighted at your success in solving the last problem, you reward yourself by buying a new Lamborghini Sesto Elemento, which can accelerate from 0 to 62 mph in 2.5 seconds. (*Note:* I am not recommending you spend your money this way every time you are happy at solving a physics problem. You are more than welcome to indulge your professor with a new car, however.)
  1. What is the magnitude of the car's acceleration in SI units?

  2. The latest figures on speeding-related car accident fatalities show 32,788 deaths per year in the United States (*L.A. Times,* 2011). In order to test the accuracy of these statistics empirically like all good scientists, you accelerate from rest to 90 mph. How long will this take? (*Note:* The authors, publishers, editors, and professors all advise against this experiment, and relinquish legal responsibility for those who try it.)

3. The maximum speed of the Sesto Elemento is 180 mph, which is definitely in the illegal range on most highways. How long will it take to accelerate from rest to its maximum speed?

4. Will doubling the time always double the change in speed? Why?

I. Suppose you are standing on the edge of a cliff 400 m high, and you throw two balls in the air: one (called ball A) directly upward with a speed of 17.6 m/s, and one (called ball B) directly downward with a speed 17.6 m/s.
   1. How high will ball A rise?

   2. How long will it take ball A to reach its zenith?

3. How much time will ball A spend in the air?

4. How much time will ball B spend in the air?

5. What will be the final velocity of ball A?

6. What will be the final velocity of ball B?

J. Delighted by your success in physics, you try your hand at rocket science, hoping to get to the moon, or at least to orbit. (*Note:* Please do not try this at home. It's actually a bit dangerous.) You manage to build a rocket which, being heavy, is launched with an initial speed of only 3.00 m/s, but it is accelerating with a constant upward acceleration of 125 m/s^2. After 34 seconds, the engines burn out and the rocket is in free fall.

1. What is the maximum height your rocket will reach?

2. How long after liftoff does the rocket reach its maximum height?

3. Assume for the sake of simplicity that the acceleration due to gravity is -9.8 m/s^2 everywhere. (It will actually get closer to zero the higher the rocket gets.) What will be the velocity of the rocket just before it crashes back down to the ground?

4. How long will the rocket be in the air?

5. What is the total displacement of the rocket, assuming one-dimensional motion?

K. You and a physics friend are designing an experiment to measure the acceleration of gravity. Here's how it works: Holding a ball at rest, you drop it from a window 35 m above the ground. Your friend on the ground is holding two radar guns 1 m above the other one, and each one will detect when the ball crosses its path. (You can assume that the average speed during that 1 m interval is approximately the final speed of the ball.) Having calculated the final speed of the ball, your friend sets a pitching machine to that speed and shoots the ball upward, shooting the ball at the exact moment that you drop another ball from the same height. These balls are actually water balloons filled with dye, and when they hit each other they splat colored dye all over the wall, so you can measure how high the balls were when they hit each other.
1. Suppose the radar guns register the ball crossing their paths 0.045 seconds apart. What speed should your friend set the pitching machine to?

2. How long after you dropped the ball did the "splat" take place?

3. Suppose now you measure the "splat" at 25 m above the ground. What was the final velocity of each ball?

4. Calculate the acceleration due to gravity from your data in questions 1–3.

5. Now suppose you want to check your answer against the experimental value for the acceleration due to gravity, -9.8 m/s^2. What speed *should* the radar guns be registering the final velocity of the ball as being?

6. Given this speed, what time interval apart should they be registering the events?

7. Given this speed, how long after the machine shoots the ball upward should the splat take place?

8. How high above the ground should the splat take place?

Two-Dimensional Kinematics 4

In Chapter 2, we briefly introduced the concept of vectors. A vector is a mathematical object with a magnitude and a direction; it can be represented graphically by an arrow starting at the origin and ending at some point on a Cartesian coordinate system. The length of the arrow is the magnitude of the vector; the angle from the x-axis rotating counterclockwise to the arrow is the direction; the x-coordinate of the point where the vector ends is the length of the x-component of the vector, and the y-coordinate of the end point is the length of the y-component of the vector. The x- and y-components themselves are vectors starting at the origin and lying along the x and y axes.

Many of the concepts we've dealt with already—displacement, velocity, and acceleration—are vectors. We noted that the four kinematics equations we developed were vector equations, but that for the sake of simplicity, we could treat them all as scalars—and that this is okay in one dimension. Mathematically, it's not quite rigorous though, and we can't do that for two-dimensional problems.

Adding vectors and taking their dot products as required by the equations is mathematically cumbersome, however, and we won't actually do that until we absolutely *have* to—when we start talking about electric forces and fields. As a rule of thumb, it's always best to do something simple instead of something difficult when you can get the same answer, and that's the strategy we'll employ in this chapter.

The four equations we developed apply to *any* vectors which physically represent displacement, velocity, and acceleration—and since the components of vectors are vectors themselves, the equations still hold true for the *components* of displacement, velocity, and acceleration. So our strategy for solving problems in two dimensions is to break those three vectors up into their components and treat the x direction separately from the y direction. In other words, we'll have *two* sets of the four equations to deal with—one for the x-components and one for the y-components:

$$v_x = v_{0x} + a_x t \tag{4-1}$$

$$\Delta x = \frac{1}{2}(v_x + v_{0x})t \tag{4-2}$$

$$x = x_0 + v_{0x}t + \frac{1}{2}a_x t^2 \tag{4-3}$$

$$v_x^2 = v_{0x}^2 + 2a_x \Delta x \tag{4-4}$$

$$v_y = v_{0y} + a_y t \tag{4-5}$$

$$\Delta y = \frac{1}{2}(v_y + v_{0y})t \tag{4-6}$$

$$y = y_0 + v_{0y}t + \frac{1}{2}a_y t^2 \tag{4-7}$$

$$v_y^2 = v_{0y}^2 + 2a_y \Delta y \tag{4-8}$$

We could replace the x's with y's with impunity because both are symbols for position—the equations don't care whether the displacement is along the x or y direction, because any choice of a coordinate system is arbitrary. There isn't any "natural" up or down as was mistakenly thought by Aristotle; we chose more or less arbitrarily what was x and what was y.

You may wonder why we used the components in the fourth equation, since that was already a scalar equation. The reason is that the products in the fourth equation were *dot products*, not the products of two scalars. When the vectors aren't pointed in the same direction, you can't just take the magnitudes and multiply them together. (Actually, what you can do is take the magnitudes and multiply them together with the cosine of the angle between them—but this is easier.) And what we're interested in isn't really v^2 or $\mathbf{a}*\mathbf{\Delta x}$, rather v, a, and Δx—and taking the square root of the dot product v^2 isn't going to give the magnitude of v or of either component. Mathematically, this is a recipe for disaster, a boiling kettle of errors waiting to happen. Separating the components is the best way to make sure everything is done correctly.

As always, the components of any vector **A** with magnitude A and direction θ are separated as follows:

$$A_x = A \cos \theta \tag{4-9}$$
$$A_y = A \sin \theta \tag{4-10}$$

Once one has separated the components from each other, the only variable held in common between the two is the one scalar quantity found in them, time. There is no foolproof algorithm guaranteed to show how every problem can be solved, but *in general*, a problem is going to be giving information about the y-components of motion and asking for something about the x-components of motion. Time is going to be the quantity you will have to solve for when going from the y-components to the x-components, because it's the only thing the equations have in common.

Usually, a two-dimensional problem will ask a bunch of questions that only pertain to the y-components of motion—how high does a projectile go, how much time does it spend in the air, and so on—and then ask for the *range* of the projectile, which is the x-component of its displacement. Most often, we will ignore air resistance for simplicity's sake, in which case there will be no x-component to the acceleration. For problems in which this is true, the last step will be to use the equation

$$\Delta x = v_x t \tag{4-11}$$

which can be obtained from either the second or fourth kinematics equation with zero acceleration.

Let's illustrate how this is done with two examples.

Example 4-1. An airplane moving east at a speed of 2500 m/s at a height of 30,000 m drops a care package to a village below. Assuming that there is no air resistance, how far east will the package go before it hits the ground?

Since the x- and y-components of motion are independent of each other (they have their own equations separate from each other), and assuming that there is no air resistance (an assumption always made throughout introductory physics classes), the package will keep moving at v_x = 2500 m/s until it is stopped when it hits the ground. We just need to know how long that takes.

As for the y-components, it will start with v_{0y} = 0 (as before, it is dropped, not thrown), and fall with gravitational acceleration: a_y = -9.8 m/s². The displacement Δy = 2500 m, and since we are looking for time and no mention was made of final velocity, we have to use equation 4-7.

$$y = y_0 + v_{0y}t + \tfrac{1}{2}a_y t^2$$
$$\Delta y = 0t + 0.5(-9.8)t^2$$
$$-30{,}000 = -4.9t^2$$

(Note the displacement is negative, because it is the final position—the ground, 0—minus initial position.)

$$t^2 = 6122$$
$$t = 78.2 \text{ sec}$$

Now we plug this into an equation for motion in the x direction. There is no acceleration in the x direction, so we cannot use one of the kinematics equations (all of which have a in them). Instead, we'll use the definition of v_x:

$$v_x = \frac{\Delta x}{t}$$
$$\Delta x = v_x t$$
$$\Delta x = 2500(78.2) = 1.96 * 10^5 \text{ m.}$$

Example 4-2. A cannon fires a cannonball at an initial speed 150 m/s at an angle 40° above the horizon. Find (a) the height the cannonball reaches, (b) the time it takes to reach its height, and (c) the range of the cannonball.

a. The height it reaches, just as in one dimension, will be when its velocity in the y direction is 0, we use equation 4-8, but only employing the y-components of its total motion.

$$v_y^2 = v_{0y}^2 + 2a_y \Delta y$$
$$0 = (150 \sin 40°)^2 + 2(-9.8)\Delta y$$
$$0 = 9.30 * 10^3 - 19.6 \Delta y$$
$$19.6 \Delta y = 9.30 * 10^3$$
$$\Delta y = 474 \text{ m}$$

b. The time it takes to reach its height is found the same way it was in one dimension, using equation 4-5 in the y direction for the easiest method.

$$v_y = v_{0y} + a_y t$$
$$0 = 150 \sin 40° + -9.8$$
$$9.8t = 150 \sin 40°$$
$$9.8t = 96.4$$
$$t = 96.4/9.8$$
$$t = 9.84 \text{ sec}$$

c. Its range is defined as the distance it goes in the x direction. In order to find the distance it goes in the x direction, we need to find the time it spends in the air, not just getting up to its height. It will land on the ground at the same level that it started at (since no mention was made of the cannon being on a hill or in a ditch), so we need to find the time when the cannonball returns to its original height—in other words, when $y = y_f$. We're looking for time and no mention was made of final velocity, so we use the horizontal version of equation 4-7, with y and y_0 both equal to 0.

$$y = y_0 + v_{0y}t + \tfrac{1}{2}a_y t^2$$
$$0 = 0 + (150 \sin 40°)t + 0.5(-9.8)t^2$$

This equation has two solutions, as all quadratic equations do. One of them is 0, since t appears in every term and $0 = $ (anything) * t. That's when the cannonball is launched, however, and we're interested in the other solution. So we can divide every term by t, something only mathematically permitted if $t \neq 0$.

$$0 = 150 \sin 40° + 0.5(-9.8)t$$
$$0.5(9.8)t = 150 \sin 40°$$
$$4.9t = 150 \sin 40°$$
$$4.9t = 96.4$$
$$t = 96.4/4.9$$

Notice that t will be exactly twice what it was before, since 4.9 (which is in the denominator) is half of 9.8. It takes the same amount of time for a free projectile to fall as to rise, so the total time will be twice the time it takes to rise.

$$t = 19.7 \text{ sec}$$

Now that we've found t, we plug it into $\Delta x = v_x t$, which will usually be the last step for all of these problems.

$$\Delta x = (150 \cos 40°)(19.7) = 2261 \text{ m}.$$

HOMEWORK FOR CHAPTER 4

Name _____

A. A ball is thrown upward in *two* dimensions with an initial speed of 17 zlugs per megahandles, near the surface of the planet Earth. What will be the *y*-component of the ball's velocity at the peak of its trajectory?

B. Consider the projectile motion of a ball in two dimensions.

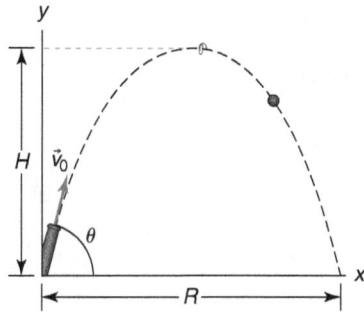

1. Where if anywhere is the speed 0? Mark the point and label it clearly on the graph.

2. Where if anywhere is the acceleration 0? Mark the point and label it clearly on the graph.

3. Where if anywhere is $v_x = 0$? Mark the point and label it clearly on the graph.

4. Where if anywhere is $v_y = 0$? Mark the point and label it clearly on the graph.

5. How does the speed of the ball change with time? Circle the correct answer
   a. It increases.
   b. It decreases.
   c. It stays the same.
   d. It varies, increasing on its way up and decreasing on its way down.
   e. It varies, decreasing on its way up and increasing on its way down.
   f. Not enough information to answer this question.

76　Foundations & Principles of Physics

6. Which scenario is the correct figure for the velocity vectors? Circle the correct answer:
   i.　0 at the top:

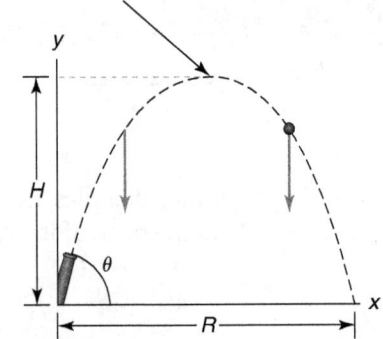

   ii.

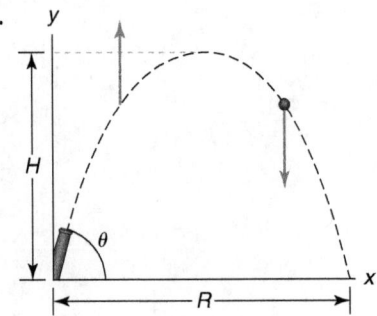

   iii.

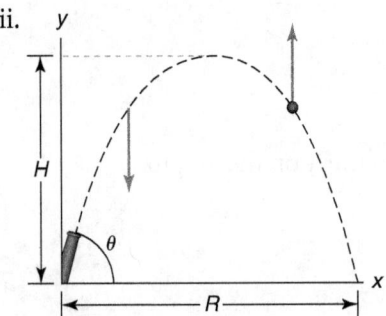

   iv.

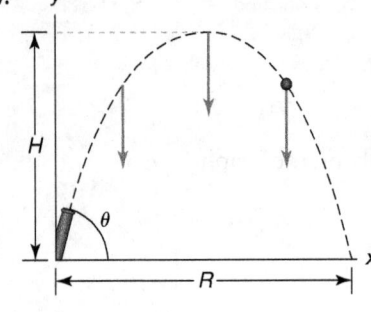

   v.

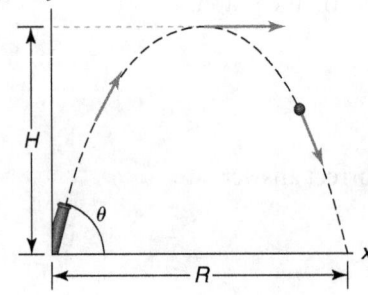

7. Which scenario is the correct figure for the acceleration vectors? Circle the correct answer.

   i. 0 at the top:
   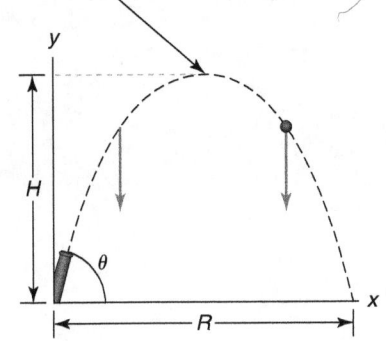

   ii.
   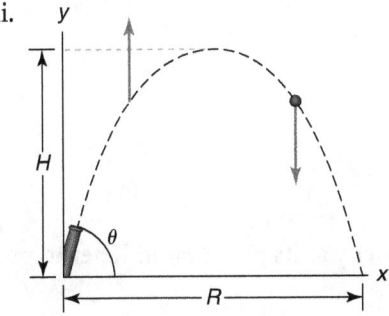

   iii.
   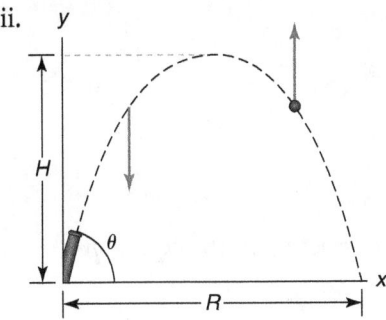

   iv.
   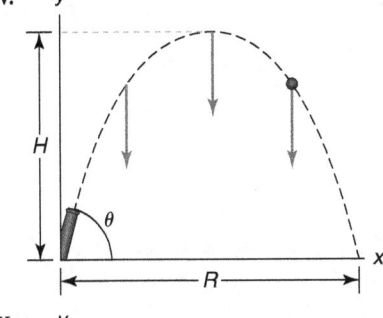

   v.
   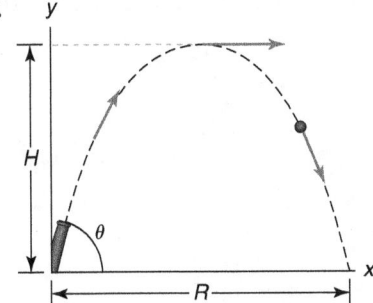

C. A cannon shoots a cannonball with an initial speed of 25 m/s at an angle 13°. (It's not a very good cannon.)
   1. What is the x-component of the cannonball's initial velocity, v_{0x}?

   2. What is the y-component of the cannonball's initial velocity, v_{0y}?

   3. What will be v_y, the y-component of the cannonball's final velocity at its peak? *Hint:* Refer to problem A.

   4. How high does the cannonball go? *Hint:* Make sure to use the correct v_{0y} in the equation $v_y^2 = v_{0y}^2 + 2a_y \Delta y$.

   5. How long will it take the cannonball to reach its peak? *Hint:* Make sure to use the correct v_{0y} in the equation $v_y = v_{0y} + a_y t$, or whichever other equation you choose to use (equations 4-6 and 4-7 will work as well—if you want to use a quadratic equation when an easy linear one will suffice, then be my guest).

6. How long will it take the cannonball to fall back down to the ground again? *Hint:* All free projectiles take the same time coming down to the level they started at as they took to go up.

7. What will be the *total* time the cannonball spends in the air? *Hint:* The total time it spends in the air is composed of the time it took to reach its peak plus the time it took to fall back down to the ground again.

8. What will be the *y*-component of the final velocity of the cannonball right before it hits the ground again?

9. What will be the *x*-component of the final velocity of the cannonball right before it hits the ground again? *Hint:* There is no acceleration in the *x* direction, so how will the *x*-component of the final velocity be related to the *x*-component of the initial velocity?

10. What will be the range (horizontal distance traveled) of the cannonball? *Hint:* Make sure you use the correct v_x in the equation $\Delta x = v_x t$.

D. A cannon fires a free projectile (ordinarily called a cannonball) at an angle 85° from the ground. The cannonball has a speed 150 m/s, and it is launched from a hill 20 m above the plain where it will land.

1. What is the x-component of the cannonball's initial velocity? *Hint:* $v_{0x} = v \cos \theta$.

2. What is the y-component of the cannonball's initial velocity? *Hint:* $v_{0y} = v \sin \theta$.

3. What is the x-component of the cannonball's velocity at the peak of its trajectory? *Hint:* $a_x = 0$.

4. What is the y-component of the cannonball's velocity at the peak of its trajectory?

5. What is the cannonball's acceleration at the peak of its trajectory?

6. What is the height **above the plain** of the cannonball at the peak of its trajectory? *Hint:* Find this using a one-dimensional equation involving v_{0y}.

7. How long will the cannonball take to reach the peak of its trajectory? *Hint:* This question also uses a one-dimensional equation involving v_{0y}. Be careful what values you give for x_f and x_0.

8. How long will the cannonball take to fall from the ground from the peak? *Hint:* Be careful what values you give for x_f and x_0.

9. What is the total time the cannonball will spend in the air? *Hint:* It goes up and then comes down. Nothing else happens. We already know the time spent going up and the time spent coming down.

10. What will be the *y*-component of the final velocity of the cannonball when it hits the ground? *Hint:* It is not landing at the same level at which it started.

11. What is the **total magnitude** of the final velocity of the cannonball when it hits the ground? *Hint:* Velocity is a vector, and the total magnitude is the hypotenuse of its *x*- and *y*-components. Use the Pythagorean theorem.

12. What is the range of the cannonball? *Hints:* (a) "Range" means how far it will go in the *x* direction. This is also a one-dimensional equation, but it involves v_{0x} rather than v_{0y}. (b) $a_x = 0$. (c) Be careful what value you use for time.

13. At what angle will the cannonball hit the ground? *Hint:* $\theta = \tan^{-1}\dfrac{v_{fy}}{v_x}$.

E. Suppose you are a general in the Napoleonic wars. You are on top of a plain 50 m high overlooking the enemy soldiers. You now have a brand-new cannon that will decimate the enemy. Unfortunately in the christening process you used some rather cheap-grade champagne, and all the water in the champagne rusted the hinges of the cannon so that it is stuck angled upward at an angle of 30° from the horizontal. Your trusty new cannon expert insisted that the ball will reach a maximum height of 150 m from the ground, but unfortunately he was killed by enemy gunfire before he could tell you where the cannonball would land on the plain below. You have soldiers on the ground and you want to make sure that only the enemy soldiers are killed by your uber-cannon. Determine the range of the cannon in order to keep your soldiers safe.

Illustration courtesy of Katherine Broberg.

F. You are once again a general in the Napoleonic wars. The battle has taken you to a trench 7 m deep. You have a mortar that will fire an explosive at a speed of 200 m/s at an angle of 80°. You have a low-flying reconnaissance airplane scouting out the enemy troops at a height of 500 m. Your pilot does not want to fly in the path of your bomb. What distance from your camp would your pilot be well-advised to avoid?

G. You are a general in the Napoleonic Wars. An enemy fort 10 m tall is located 1000 m from your cannon. Your cannon expert is insisting that you will get the best results orienting it at an angle of 25° above the horizontal. Your cannon is located on a hill 10 m tall. You have soldiers storming the plain in front of the fort in order to storm the building once the wall has been breached, and there are civilians inside that you do not want to kill. If the initial speed of the cannonball is 112 m/s, will the cannonball strike the building (at what height?), overshoot and kill the civilians inside, or hit your soldiers who are storming the plain in front of the building?

Illustration courtesy of Katherine Broberg.

H. You are a general in the Napoleonic Wars. You are on top of a plain 12 m tall, and you have a cannon that can shoot a cannonball out at 350 m/s. 800 m away is a thick forest that you cannot see into, but one of your scouts in a hot-air balloon has just signaled to you that an enemy cavalry charge inside the forest is rushing toward you at a speed of 7 m/s, and they are 1000 m from the edge of the forest. You want their entry onto the battleplain to be welcomed with a nice big fat cannonball. At what angle must you tilt your cannon so the cannonball hits the edge of the forest, and what time (starting with $t = 0$ sec at the present) must you fire it so that it hits the enemy just as they reach the forest's edge?

Forces and Newton's Laws 5

In the first four chapters, we've worked with only two fundamental concepts—position and time—and derived *all* of kinematics and solved some pretty intricate problems *only* using those two basic concepts, as well as some composite concepts whose definitions relied only on them (namely velocity and acceleration). In theory, although it would have been cumbersome and useless (since collecting like terms would cause them to cancel out leaving us with the uninteresting statement x = x for all four of the kinematics equations), we could substitute the definitions of velocity and acceleration wherever they appear in the kinematics equations and write them using position and time alone. These equations are a priori synthetic statements using only two definitions and the axioms of algebra.

In this chapter, we're going to introduce one more fundamental concept that will allow us to derive all the physics we're going to use (with the help of some more composite concepts) up until Chapter 9. That fundamental concept is *force*.

For philosophical reasons, we're going to leave "force" undefined (at least mathematically speaking) and taken as being a common fact of everyday experience. Qualitatively speaking, force can be defined as *the efficient cause (or agent) of change on an object,* or less precisely that which makes something happen. I deliberately used the old term "efficient cause" from Aristotle's *Physica* as a synonym for "agent" (a term preferred by more modern analytic philosophy) in order to point our discussion back to Aristotle's naïve and apparently commonsensical—but incorrect—ideas about inertia and force which unfortunately are the *de facto* manner of thinking about reality in all minds not trained in physics. I also kept the definition vague enough to highlight two facts about physics—(a) force is "whatever" causes the change in an object; it is a term that *references* the cause without saying *what* the cause is. In other words, while we are measuring properties of matter mathematically, we are not saying anything about what premodern philosophy called the "essence" of matter. Aristotle's statement that matter is unknowable remains true for classical Newtonian physics—all we can talk about are measurable *phenomena* about the matter. The second fact is that (b) what the aforementioned Immanuel Kant called the "noumenon" or thing-in-itself (contrasted with the "phenomenon" which we have been calling the "observable") is not under discussion, remaining consistent with either Kantian philosophy, which declared the noumenon unknowable, or with medieval realist philosophy, which declared the noumenon identical to the phenomenon. David Hume's claim that causality itself was a meaningless term is also consistent with this definition; there need not be any *reason* or *explanation* of the force in order for us to be able to measure the force. Those philosophical debates are ones that we are simply not entering, and which remain open questions given our physics—and we've been careful to make sure that our definitions are consistent with *all* of the three major philosophical interpretations of the noumenon and phenomenon (realism, skepticism, and critical conceptualism).

Now that we've clarified our stance vis-à-vis the philosophical questions that most of the earliest scientists—including Galileo, Descartes, Newton, and Kant—spilled more ink over than they spilled over mathematical physics, we can go on to make two more points discovered by Isaac Newton, which are not necessarily obvious to the naïve, untrained mind, and which contradict the physics of all pre-Newtonian philosophy.

The first point regards what is called "inertial mass," the resistance of an object to a force acting on it (proportional to the amount of "stuff" in it, or to how big and dense and bulky it is, loosely speaking). Although they had not conceived the problem in mathematical terms, pre-Newtonian physics thought of inertial mass as being defined as

$$m \equiv \frac{F}{v} \tag{5-1}$$

where F is the force and v is the velocity. Of course, one can *define* terms however one likes, and even derive a consistent system of physics from this definition—as has been done in contemporary times by some maverick physicists, notably John Ralston at the University of Kansas.[1] *However*, this definition is problematic, because given any single object with an unchanging "amount of stuff" in it—and we usually think of inertial mass as *also* being a measurement of the "amount of stuff" in something—the force will *not* be proportional to the object's velocity as this equation seems to state. Multiplying both sides of this equation by v gives us the (incorrect) relation

$$\boldsymbol{F} = m\boldsymbol{v}$$

which *seems* to correspond to everyday experience. The harder you push on something, the faster it goes; stop pushing, and it slows to a stop, sometimes stopping instantaneously. Aristotle regarded this equation as being true, forcing him to conclude that when an object is thrown the air keeps pushing on it to keep it moving. Even later commentators who improved on his physics, such as the fourth-century Christian commentator John Philoponus, who said that some intrinsic principle rather than the force of the nearby air pushing on the object is what keeps it in motion, still thought of force as being somehow proportional to velocity (even if they were inconsistent in explaining how this could be). And everyday experience seems to justify this—stop pushing on an object, and it stops moving.

The reason an object in motion stops moving is actually because another force—friction—is acting *against* the force originally applied to it, causing it to slow down and come to a stop over a nonzero quantity of time. In other words, friction causes a change in velocity, slowing down rather than speeding up, over a given time interval. Change in velocity per time has already been defined as acceleration, and recognizing this, Sir Isaac Newton explained the phenomenon of objects remaining in motion for a time after the initial force has ceased by redefining inertial mass as

$$m \equiv \frac{F}{a} \tag{5-2}$$

or to give the more familiar form of this equation,

$$\boldsymbol{F} = m\boldsymbol{a} \tag{5-3}$$

Note that because acceleration is a vector, force also is a vector (a point obscured by our choice to define the scalars first–it can be proven with a little more work, however, that $\boldsymbol{F} = m\boldsymbol{a}$ is a true equation). A force is always applied on an object in a certain direction; you apply a force for example when you push on a heavy crate, and there's a certain magnitude to how hard you push on it and

---

[1] Demonstrated in a personal conversation with the author, who unfortunately had to decline a research assistantship under him while in graduate school. Ralston provides compelling arguments for the merits of using this definition of inertial mass, which are beyond the scope of this book.

you push on it in some direction as opposed to a different one. Force is actually a good example of a vector because it brings the concept of vector down into the world of everyday life in a very tangible and relatable way.

This equation is a mathematical formulation of the well-known but little-experienced "principle of inertia": "an object in motion stays in motion, and an object at rest stays at rest." This is consistent with and in fact implied by Galileo's principle of relativity, which says that the kinematics equations (which he derived) hold true from *whatever reference frame they are measured at*, even ones themselves in uniform motion relative to someone else—in other words, there is *no mathematical difference* between rest and uniform (acceleration-free) motion except the viewpoint of the person measuring it. As said above, this principle seems contrary to experience because of the ubiquitous presence of friction. An object in motion tends to come to a stop (due to friction), not stay in motion. The almost-frictionless environment of outer space is the only place where Newtonian physics and everyday experience unite in a way that is clearly obvious to the nonphysicist.

One of the implications of the principle of inertia is that the word *motion* properly applies to *acceleration*, not to velocity. Something moving with uniform velocity is at rest when measured from its own reference frame; a person sitting in a train moving at 70 km/hr can walk around and balance a cup of coffee just as easily as he could sitting in his kitchen, setting aside the vertical motion of the train bumping up and down. Aristotle's *Physica* and later metaphysical reflections on the existence of God are permeated with discussion of "motion" by which he simply meant "change" (in qualitative properties as well as quantitative ones); the only real change that occurs in an object is when it accelerates. An object moving at uniform velocity is not changing; only its position along the arbitrary coordinate system we impose on it is changing. The word *motion* as found in ancient, medieval, and modern neo-Thomist philosophical treatises is consequently used incorrectly, *pace* Ralston.

The second point that Newton discovered is that gravity is a force. This *is* obvious to us today—jump out of an airplane and there is obviously a change in motion causing you to plummet to the ground where you will provide a lesson in anatomy and jello dynamics to the bystanders, at the cost of your own comfort (and life)—but under the influence of Aristotle pre-Newtonian thinkers had thought that things had a "natural motion" proper to their "essences," which caused them to move either in an absolute "down" (if they had mass) or an absolute "up" (in the case of fire). Unquantifiable principles like "essences" and "forms" which populate the Aristotelian universe are not employed by Newtonian and post-Newtonian physics. Things fall down because a force—gravitational force—is acting on them.

The reason this is counterintuitive is because medieval and ancient philosophers beginning with Aristotle had thought that the *speed* or velocity at which a falling object hit the ground was proportional to its weight. If you drop a boulder and a feather from a tower at the same time, one would always expect the boulder to come crashing to the ground quickly while the feather gently floats its way down. The actual reason for this is that another force—air resistance—acts stronger on the feather than on the boulder, slowing down its acceleration. Drop both from a tower on the moon or in a vacuum chamber on earth in which all the air has been removed (an easy experiment to do today), and one will find that both the boulder and the feather hit the ground at *exactly the same time*. This was allegedly first demonstrated by Galileo in a probably apocryphal story in which he dropped two cannonballs of unequal weight from the Leaning Tower of Pisa, and contrary to the predictions of Aristotle and 2,000 years of otherwise remarkably intelligent thinkers who had never thought to try this, they landed at the same time. In order for them to land at the same time given the same displacement and given that they both were dropped from rest, the second kinematics equation ($\Delta x = \frac{1}{2}(v + v_0)t$) tells us that the final velocities had to be the same—and if the initial and final velocities were the same given the same time interval, the definition of acceleration tells us that they had the same acceleration. Weight, in other words, is nothing other than *gravitational force*, proportional to the amount of stuff in the object—which ever since Newton we have been able to call either inertial mass or gravitational mass; they are the same thing since gravity is a force—but with the *same acceleration* for all objects regardless of their mass.

This of course all assumes that mass does not change—an assumption that would have to be thrown out to keep Aristotelian physics consistent—and this assumption is based on our intuitive understanding of mass as the measurement of the amount of "stuff" in a substance. This understanding is so deeply ingrained in our consciousness that when the Nobel laureate Richard Feynman, possibly the greatest and certainly one of the greatest physicists of the last century, tried to solve certain problems in special relativity by introducing a variable "relativistic mass" that depends on the particle's velocity, it was almost universally rejected by the scientific community, not because it's wrong (anything self-consistent with its own definitions is correct) but because that's just not how we think of mass as a basic concept.

Legend tells that Newton discovered this insight seeing an apple fall from a tree, or in a more developed version of the story, fall onto his head from a tree. However he actually came about it, the insight enabled him to develop the first equation allowing us to calculate the magnitude of a force given other measurable facts about the objects involved. His equation, called Newton's universal law of gravitation, is written

$$\boldsymbol{F}_g = G \frac{M_1 m_2}{r^2} \hat{r} \qquad (5\text{-}4)$$

This is a force *between* two objects—the earth, and the apple that bonked his head—and Newton reasoned that the force is proportional to both of the masses involved. (Conventionally the mass of the larger object, in this case the earth, is denoted with a capital "M," and the mass of the smaller object with a lowercase "m.") He also reasoned that if one of the objects (say, the bigger one) were regarded as being at the center of a sphere, all of the smaller objects on the surface of the sphere would fall to the bigger one at the same rate—Galileo's cannonball experiment. Newton was the inventor of calculus, and he argued that summing up or integrating over all of the force on the surface of the sphere gives us an *invariant quantity*. If you view the surface of the sphere as a rubber ball the tautness or density of which gives us the magnitude of the force, the total amount of rubber remains unchanged no matter how big we blow that sphere up to be. It will be denser (giving a stronger force) when the ball is smaller, and more taut and less dense (giving a weaker force) when the ball is bigger. The density of the rubber is inversely proportional to the surface area of the ball, and the surface area of a spherical surface is given by $A = 4\pi r^2$ from basic geometry. So the gravitational force will be proportionate to $\frac{1}{r^2}$. Newton was only thinking about quantities being proportional to each other, and he didn't know what constant to include to make the proportionality relation an equality, so he wrote that proportionality constant—including the 4π from the surface area of the sphere—as a capital G. Today G has been measured to be $6.673*10^{-11}$ N m^2/kg^2. (N stands for the units of force, or kg m/s^2. In honor of the discoverer of the universal law of gravitation, in the SI system kg m/s^2 has been appropriately renamed by the shorthand "newtons" or N.)

Force is a vector because acceleration is a vector and force is a scalar quantity (mass) times acceleration, yet mass and r^2 (the dot product **r** * **r**, if you want to be mathematically rigorous about it) are both scalars. The bold $\hat{r}$ to the right of all the other quantities in the equation is called a unit vector, a vector with length 1 pointing in the direction r, or the line of sight between the two objects attracting each other gravitationally. (Obviously this direction changes when an object in orbit changes, but that's okay—it's technically not a Cartesian direction at all but a direction in *spherical polar coordinates,* which is all you need to know about them right now.) The "r" in r^2 does **not** have unit length; it is the distance between the two objects that are attracting each other. When the earth is one of those objects, that's actually the distance from the falling object to the *center* of the earth, since that's where the center of mass is—a subtlety to be introduced in due time. A gravitational force is still acting on us when we are sitting in our lawn chair at rest, according to the universal law of gravitational; if r^2 were from the surface of the earth rather than the center, our feet on the ground would experience an *infinite* force (since the force would be proportional to some numbers divided by 0) and like elephants we would find it impossible to jump. We'd even find it impossible to lift our feet, and our whole bodies would come crashing down pulled by our feet, something that in real life only happens in the unfortunate circumstance of entering a black hole. Instead, when using Newton's universal law of gravitation to find the force between the earth and the person on the surface, we treat the whole earth as being a point in which all the mass is concentrated (that point being called the center of mass, located near the center of the earth).

If a force is acting on us, and it acts in the direction of the center of the earth, and if **F** = m**a**, why then don't we crash through the ground accelerating toward the center of the earth? The simple reason is that the ground holds us up, exerting a force of its own that balances the force of the earth. That force is called the **normal force** because it always acts perpendicular to the surface of the earth, and the word *normal* comes from the Greek word for "perpendicular." Why the word *normal* means what it does in everyday language is a mystery to physicists, who are generally unfamiliar with the latter meaning of the word.

If **F** = m**a** and $F_g = G\frac{M_1 m_2}{r^2}\hat{r}$, then for a gravitational force $m_2 \boldsymbol{a} = G\frac{M_1 m_2}{r^2}\hat{r}$ and dividing by m_2 gives us $\boldsymbol{a} = G\frac{M_1}{r^2}\hat{r}$ the magnitude of which *near the surface of the earth* is conventionally labeled by a lowercase g and happens to be equal to 9.8 m/s². The gravitational force acting on an object always pulls it "down"—that's the definition of the word *down*—so using the symbol w for weight/gravitational force, we have $w = -mg$. This is the equation that should be used for the gravitational force on earth near sea level; for any other situation, the acceleration will be different than 9.8 m/s² and the universal law of gravitation must be used if the acceleration is not given.

Since the gravitational acceleration in general is $\boldsymbol{a} = G\frac{M}{r^2}\hat{r}$, and we know the gravitational acceleration on the surface is 9.8 m/s², if we know the mass and radius of a planet relative to that of the earth, we can find the gravitational acceleration on that planet relative to 9.8 m/s². The ratio between the two accelerations is $\frac{a}{9.8}$, which can be written using Newton's Universal Law of Gravitation as

$$\frac{a_{planet}}{9.8} = \frac{G\dfrac{M_{planet}}{r^2_{planet}}}{G\dfrac{M_{earth}}{r^2_{planet}}} = G\frac{M_{planet}}{r^2_{planet}} * \frac{r^2_{earth}}{GM_{earth}} = \frac{M_{planet}}{M_{earth}} * \frac{r^2_{earth}}{r^2_{planet}} \quad (5\text{-}5)$$

Example 5-1

For example, suppose we have a planet seven times the mass of the earth, with a radius four times the earth's average equatorial radius. While we could laboriously calculate that planet's mass in kilograms and radius in meters and use Newton's universal law of gravitation to find its gravitational acceleration, equation 5-5 is much faster. The ratio $\frac{M_{planet}}{M_{earth}}$ is simply 7, and the ratio $\frac{r^2_{earth}}{r^2_{planet}}$ is simply $\left(\frac{1}{4}\right)^2 = \frac{1}{16}$. So $\frac{a_{planet}}{9.8} = 7 * \frac{1}{16}$, or $a_{planet} = 9.8 * \frac{7}{16} = 4.29\frac{m}{s^2}$. (5-6)

So far we haven't mentioned the three most important laws Newton discovered about forces in general, just some reflections on his starting point, a consideration of gravity as being a force. Let's step back for a moment and consider the fundamental concepts we based our kinematics on—displacement, velocity, and acceleration. If multiple displacements occur, the total displacement is the vector sum of all the individual displacements. In other words, if I walk 3 miles east, 4 miles north, 3 miles west, and then 2 miles south, I'd end up in the same place as if I'd just walked 2 miles north. This is true of all vectors, and all the concepts that have been introduced have been derivative of position, time, and force and therefore have been scalars or vectors (for which addition is defined) as well. Advanced mathematics courses in "abstract algebra" or "modern algebra" may introduce mathematical objects which you *can't* add in any way (for example, multiplicative groups); the good news is that at least until we get to quantum field theory (which the users of this textbook won't see for a few more years) we won't have to worry about these. If Suzy pushes on a heavy crate with a force 5 N and Bobby pushes back on the other side of the crate with a force −5 N (which we call negative only because it acts in the opposite direction), the total force on the crate will be 0 and it will not move.

Once again, this fact is not as obvious as one may initially think. It implies that forces are not *things*. Two equal and opposite forces can cancel each other out to give a net force and net acceleration of 0, but you can't add two bananas pointing in opposite directions and find zero bananas. Bananas are things, not vectors, and if one were to make an analogy between bananas and mathematical objects they would be more like scalars (since one banana + one banana = two bananas, regardless of which direction they are pointing—you will never see one banana + one banana equal anything other than two bananas, unless you are counting in binary numbers which is a modular arithmetic group, and we've stated already that we're not going to deal with any groups except the sets of real scalars and vectors. Hopefully this will not disappoint too many computer science majors who might prefer to think in binary.)

The difference, although presented here in a rather crude manner, is actually quite subtle. If the total quantity of something can be found by adding together the component quantities in such a way as to permit them possibly canceling out, we say that it obeys the *principle of superposition*. The most obvious example of this would be waves—add two waves together that are shifted by half a wavelength and they will cancel out. Forces are the same way. Suzy's force pushing the box to the left cancels out Bobby's force pushing the box to the right. It turns out in subatomic physics that some particles—called *bosons* after the Indian physicist Satyanendra Bose—actually obey this principle, a good example being the photon or particle of light. More thing-like particles which do not are called *fermions* after the Italian physicist Enrico Fermi, but it turns out that the only difference between bosons and fermions is whether their spin (a measurement of how they align themselves in the presence of a magnetic field) is measured in whole integers (bosons) or half-integers (fermions). And in one of the truly bizarre phenomena that soon may permeate everyday technology, scientists have been able to couple two electrons (which are fermions) together so that they look like single particles called "Cooper pairs" with a total spin $\frac{1}{2} + \frac{1}{2} = 1$, making the electrons (which are otherwise like "things"—they have mass and are some of the fundamental constituents of atoms) superimpose on top of each other and cancel each other out. This behavior is manifested at the macroscopic level by giving rise to the phenomenon of superconductivity, which allows high-speed bullet trains to minimize friction by floating a couple centimeters above their rails.

The principle of superposition applied to forces is called Newton's first law. That's all there is to the first law—Bobby's force pushing one way is canceled out by Suzy's force pushing another way. But while it may seem trivial, it allows us to extend the definition of inertial mass into Newton's second law.

Since forces obey the principle of superposition, the total net force is the vector sum of the individual forces—that's Newton's first law again. We can write that mathematically as

$$\sum \mathbf{F} = \mathbf{F}_{tot} \tag{5-7}$$

But for any force, the definition of inertial mass tells us that $\mathbf{F} = m\mathbf{a}$. Substituting this for the total force in Newton's first law gives us Newton's second law, the law which we will actually use to solve (almost) all of our problems in this chapter:

$$\sum \mathbf{F} = m\mathbf{a} \tag{5-8}$$

All of the forces on the left-hand side are the *individual* forces that are acting on an object—weight, the normal force, the force from an engine, frictional force, the force of someone pushing on it, tension (the force applied on an object by a rope or cable), the electrical force, or anything else. The acceleration on the right-hand side is the *total* acceleration of the object. This law holds true for *all* forces, and since the vector components of forces are themselves forces (they have the same units, they are vectors, and the definition of inertial mass still holds true because force is proportional to acceleration), so this equation is most easily handled by taking the components of the vectors separately:

$$\sum F_x = ma_x \qquad (5\text{-}9)$$

$$\sum F_y = ma_y \qquad (5\text{-}10)$$

In many cases, the object in question will not be accelerating. It will be usually at rest, or possibly moving at a uniform velocity. This special case is called *equilibrium*, and is seen in many problems because it's easier to work with. In this case, since the x- and y-components of acceleration are perpendicular to each other and therefore cannot cancel each other out, both must be 0:

$$\sum F_x = 0 \qquad (5\text{-}11)$$

$$\sum F_y = 0 \qquad (5\text{-}12)$$

The way to solve problems using these equations (equilibrium or otherwise) is to simply find expressions for each one of the forces involved, find the x- and y-components of each force, add them all together, and set them equal to the mass times the component of the acceleration (0 if the object is in equilibrium). The weight is given by the formula given above for weight or by the universal law of gravitation, the frictional force is given by its own equation to be discussed next, and usually the only other forces in a problem (at least until Chapter 9) will be tension and normal force, which do not have universally applicable equations that we can use to substitute expressions for them. Instead, we leave them as unknown variables (called T and N respectively), and solve for them. We have two equations (one for each direction) and two unknowns, and that's an algebra problem.

Friction is a bit of an oddity. It's a *reactionary* force, meaning that it always makes the acceleration of something closer to 0. One is certainly a special person if they've seen friction speed something up or cause a chair to fly across the room on its own volition—it just doesn't happen. (Wheels are a different case—we're assuming throughout in this chapter that there's no *internal* motion, just rigid bodies moving forward, and this isn't the case for wheels.) So the magnitude of the frictional force is never greater than the force being applied on it.

For an object at rest, the frictional force is equal and opposite to the other forces acting on it. In other words, the object is at rest because the frictional force cancels them out. The maximum frictional force for an object at rest—the frictional force present when the object finally starts to move—is given by the following equation:

$$f_{max} = \mu_s N \qquad (5\text{-}13)$$

where N is the normal force (what keeps you from falling through the ground and accelerating at -9.8 m/s^2 toward the core of the earth while sitting in your lawn chair) and μ_s is a measured value characteristic of the surface called the "coefficient of static friction". Normal force is what usually counteracts gravitational force, so on a flat surface the normal force is equal and opposite to the weight, which makes sense—a heavier object is harder to push into motion, and that's because the frictional force acting against you is higher. On a slope, the normal force is lower because the normal force is always perpendicular to the surface—the normal force is given by $N = mg \cos \theta$ where θ is the angle of the slope. (Cos $0° = 1$, so on a flat surface $N = mg$. Cos $90° = 0$, so presuming that no extra force is pushing you into the wall, the normal force will be 0 and you will fall straight down. It's another way of saying that the side of a vertical wall won't hold you up, which is common sense.) Since the normal force is lower on a slope, the frictional force is too—that's because there's a component of the gravitational force that points in the same direction of the slope and makes something go down the slope, and that force is equal to mg sin θ. (The slope and the normal force are by definition perpendicular, so we can just tilt our x-axis to go in the direction of the slope and our y-axis in the direction of the normal force, and we have our two components of the weight that way.) It's easier to push a heavy box on a slope into motion because gravity is helping you (unless you're trying to push it uphill, in which case gravity is working against you).

Once an object is in motion, the frictional force it applies actually dips a little bit and then levels out, remaining constant no matter how much further force is applied to the object. The frictional force of a moving object (once it has leveled out) is given by the equation

$$f = \mu_k N \qquad (5\text{-}14)$$

where μ_k is called the "coefficient of kinetic friction"; the reason the frictional force applied by a moving body is a little bit less than the maximum friction applied on a body at rest is that μ_k is slightly lower than μ_s. Of course, that's just a restatement of the fact without explaining *why* this is true (which is beyond the scope of this book), but it's a useful fact to remember. Note that in all of this we haven't violated Galilean relativity; these forces are always defined from the reference frame of the surface causing the normal force, because the same surface also causes the frictional force.

The hard part is often figuring out which forces go which directions. Some steps can be given as rules to follow:

1. Write down all of the forces involved and figure out if they are affecting the object in the horizontal direction, vertical direction, or both.
2. If the object is in the air or on a flat surface, draw a plain rectangular box and for all the horizontal forces draw them as arrows coming out of the sides of the box (left side for forces pulling it backward, right side for forces pulling it forward). For all the vertical forces draw them as arrows coming out of the top and bottom of the boxes (for forces going up and down respectively). For all the forces going both ways, figure out whether these forces are applied in the northeast, northwest, southeast, or southwest directions (taking y as pure north and x as pure east), and draw the x- and y-components alongside the others taking the magnitude times cos θ for the x-component and the magnitude times sin θ for the y-component. The y direction is straight up, and the x direction is straight to the right.
3. If the object is on a slanted surface, one *can* use step 2—with the result that all of the forces except weight will have two components—but it is easier to tilt the coordinate system so that the y or "vertical" direction is perpendicular to the slope and the x or "horizontal" direction is parallel to it. That will usually make weight the only force with components. Repeat steps 1 and 2, except that now the x-component is the magnitude times the *sine* of θ and the y-component is the magnitude times the *cosine* of θ. Ask your instructor for a geometric explanation of why this is true.

The diagram thus drawn is called a *force diagram*. Writing down Newton's second law for both directions is quite easy once the force diagram has been drawn: For $\sum F_y = 0$, just add all the forces on top of the diagram and subtract the ones on the bottom, and set them all equal to 0. For $\sum F_x = 0$, add all the forces on the right of the diagram and subtract the ones on the left, setting them all equal to 0.

Most of the problems we've done so far using the kinematics equations have been "free projectile" problems, which by definition have only one force acting on them—weight. Now we'll be able to solve problems with more forces.

Let's take an example. A ski lift chair with a mass of 300 kg (including the passengers) is hanging from a cable, and it's heavy enough so that at the top of the hill it sags with the cable going upward at 25° on either side. Assuming that the chair is in equilibrium at the top of the hill, what is the total tension in each side of the cable?

Recall that tension is just the force exerted on the chair by the cables. We have three forces—the two sides of the cable, and the weight of the chair—so let's list which forces are horizontal and vertical:

Weight: Vertical (down)
Tension 1: Vertical and horizontal (up and to the right)
Tension 2: Vertical and horizontal (up and to the left)

Translate this into a force diagram:

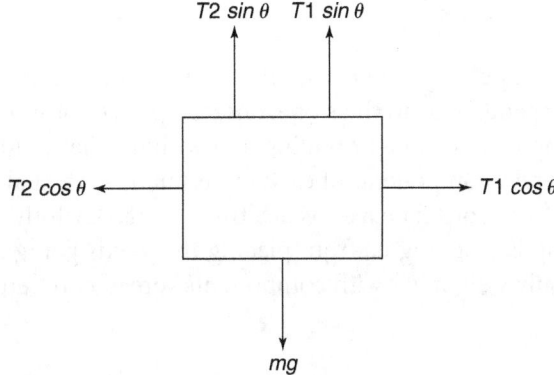

Following the rules given for converting force diagrams into Newton's second law, we add the forces on top and subtract the ones on bottom, and add the ones on left and subtract the ones on the right. They are all equal to 0, since the chair is in equilibrium.

$$T_2 \sin \theta + T_1 \sin \theta - mg = 0$$
$$T_1 \cos \theta - T_2 \cos \theta = 0$$

Let's solve this algebraically as far as possible before plugging in the numbers, which is generally a good thing to do. Note that we only used one symbol for the angle, θ, only because both ropes are tilted at the *same* angle, 25°. Otherwise T_1 and T_2 would have components based on different angles with different names.

Taking the second equation and adding the second term to the other side,

$$T_1 \cos \theta = T_2 \cos \theta$$

The cosine of 25° is not 0, so we can divide both sides by $\cos \theta$ giving us $T_1 = T_2$, which is not too much of a surprise given the symmetry of the problem.

Calling them both "T" for the sake of simplicity, the first equation becomes

$$T \sin \theta + T \sin \theta - mg = 0$$
$$2T \sin \theta = mg$$
$$T = \frac{mg}{2 \sin \theta}$$

Plugging in the numbers, $T = (300 * 9.8)/(2 \sin 25°) = 3.48 * 10^3$ N.

Example 5-2

Let's take a simple example: Pulling a 30 kg box up a 30° slope with a coefficient of kinetic friction 0.8, let's find the tension in the rope.

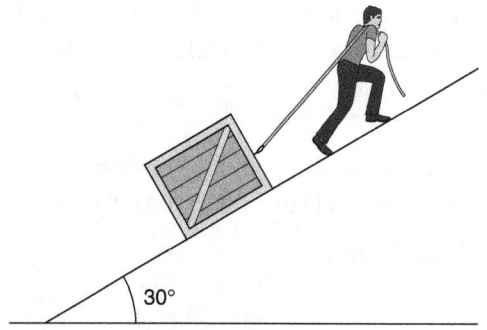

Here we have tension pulling the box uphill, friction reacting against it down the slope, normal force acting perpendicular to the slope, one component of weight acting against normal force, and one component of weight pulling it downhill. One could place coordinate axes up and down, taking x and y components of each force, but this is actually harder than we need to work. The placement of coordinate axes is arbitrary (*pace*, Aristotle, there's no *natural* "up" or "down"), and nothing is stopping us from placing the y-axis going in the direction of the normal force, leaving only weight left with components spread between two directions:

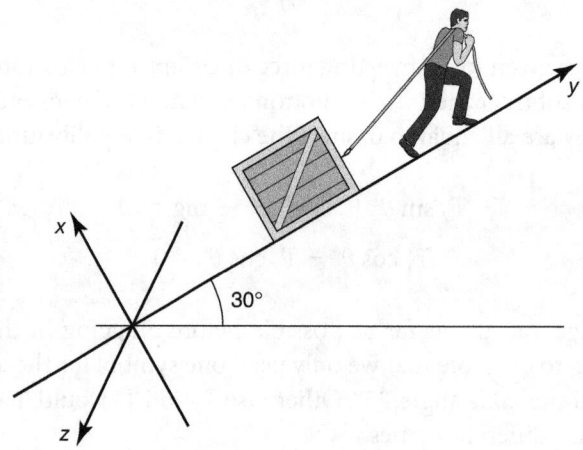

This configuration puts the normal force along the y-axis, tension along the positive x-axis (uphill), and friction along the negative x-axis (downhill). There will be both negative x and negative y components to the weight; however, we have to be a little bit careful about how we take those components. The coordinate axes are tilted, so the 30° angle *isn't* the one we're measuring the cosine and sine of the angle from anymore. A little geometry can help us, however.

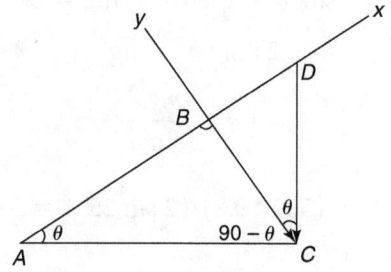

The image below shows the slope with the angle θ and the new x and y axes. Because the sum of all the angles in any triangle is 180°, and a right triangle has an angle 90°, we can label the angle across from θ as being 90° − θ. One can clearly see that the line segment AC and the weight vector mg form a right angle (90°), so the angle immediately counter-clockwise to the dashed line forming the y-axis (angle BCD) must be 90° − (90° − θ) = θ. Now the y-component of mg is the line segment BC, and by definition of cosine, cos θ = (adjacent side)/(hypotenuse) = BC/mg. Multiplying both sides by mg, $BC = mg \cos \theta$. In other words, because we have rotated the coordinate axes, the y-component of the weight is $mg \cos \theta$ now, not the x-component. And by a similar argument, the x-component of the weight (the component going downhill) will be $mg \sin \theta$.

Consequently, our force diagram will have $mg \sin \theta$ on the left, and $mg \cos \theta$ on the bottom:

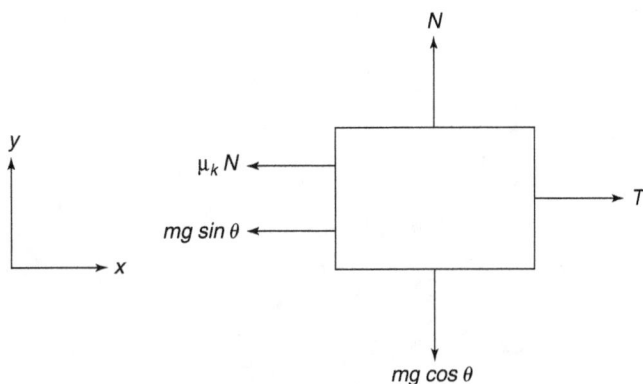

And Newton's Second Law will be written as the following pair of equations, with the acceleration being 0 for both components because we are in equilibrium:

$$N - mg \cos \theta = 0$$

$$T - \mu_k N - mg \sin \theta = 0$$

The second equation has both N and T, so we need to solve the first equation for N first:

$$N = mg \cos \theta = 30(9.8) \cos 30° = 254.6 N$$

Plugging this into the second equation,

$$T = \mu_k N + mg \sin \theta = 0.8(254.6) + 30(9.8) \sin 30° = 350.7 N$$

We can't close this chapter without mentioning Newton's third law. Popularly phrased, it says that *every action has an equal and opposite reaction*. A more sophisticated way of thinking about this is that every time one object exerts a force on another object (and in reality, every force is exerted by one object on another—at the subatomic level, we stop speaking of "forces" altogether and start speaking of "interactions" instead), the second object exerts the opposite force on the first one. Phrased another way, the force exerted by object 1 on object 2 is equal and opposite to the force exerted by object 2 on object 1, with absolutely no regard to which one is "causing" the force to be applied. (Again, we're chickening out of the philosophical debate about causality.) Mathematically, this is stated as

$$\mathbf{F}_{12} = -\mathbf{F}_{21}$$

We won't deal with this this in solving problems any further until we rewrite this law as the law of conservation of linear momentum in Chapter 7, in which respect it will be incredibly useful. However, it is enlightening to understand that this is the reason that rockets work—the force that the rocket applies to the burning fuel is equal and opposite to the force that the fuel applies to the rocket, and the rocket consequently shoots upward. It's also the reason why you can't have a tug-of-war in outer space, without friction counteracting those reaction forces.

HOMEWORK FOR CHAPTER 5

Name _____

A. You are walking out of a grocery store pulling a cart full of groceries behind you with a rope in a flat parking lot. (Don't ask me why you are pulling it with a rope instead of pushing it. The nice men in white coats will be here soon.) Name the forces keeping it in equilibrium in the y direction, and the forces keeping it in equilibrium in the x direction.

B. Suppose you have a heavy wooden crate resting on the floor. Is it easier to push it until it has a speed of 5 m/s, or to take the same crate moving at 5 m/s and push it until it has a speed of 10 m/s? Explain your answer.

C. If only one force acts on an object, can it be in equilibrium? Explain your answer.

D. A satellite with a mass of 2000 kg is orbiting the earth at a height of 250 miles.

1. Which direction does the force of gravity acting on the satellite go?

2. Which direction does the force of gravity that the satellite exerts on the earth go?

E. Your brand-new $500 physics textbook has a mass of 100 kg and is lying on your table. Assume your textbook still has a mass of 100 kg.
   1. Which direction does the force of gravity acting on the textbook go?

   2. Which direction does the normal force acting on the textbook go?

   3. What is the weight of the textbook?

   4. What is the normal force on the textbook?

102 Foundations & Principles of Physics

F. A massive earthquake rips through Columbus, leaving your desk tilted at an angle of 47° down from the floor.

   1. Draw the force of gravity of the textbook.

   2. Now draw the normal force on the textbook.

   3. What is the weight of the textbook?

   4. What is the normal force on the textbook?

G. You spend an entire afternoon repairing the damage of the earthquake, and then set your trusty physics textbook back flat on the table. You are talking to a cute girl (or boy) on the other side of the table and trying to impress her (or him) by conspicuously leaning on the textbook so that she (or he) sees what an impressive physics genius you are. You are leaning on the book with a force of 760 N. Assume the textbook still has a mass 100 kg.

  1. What is the normal force that the table is exerting on the textbook?

  2. What is the difference between the normal force now and the normal force when you were not leaning on it?

H. You are spending your Christmas break in Minnesota, absolutely the best place in the world to spend a winter. After some light flurries deposit 28 feet of snow on the ground, you take advantage of the situation to go sledding—and to solve physics problems. A wooden toboggan pulled across wet snow has a coefficient of kinetic friction $\mu_k = 0.1$ (Serway and Jewett, *Physics for Scientists and Engineers*, 9th ed., Table 5.1).
   1. You are walking up a gentle snowy slope of 87° dragging a 30-kg toboggan.
      a. Draw a force diagram of the toboggan. *Hint:* This is easier done with the *x*-axis defined in the direction of the slope.

      b. Use the force diagram to write Newton's Second Law in the *x*-direction $\left(\sum F_x = 0\right)$ for the toboggan (writing out expressions or symbols for each individual force).

c. Use the force diagram to write Newton's Second Law in the y-direction $\left(\sum F_y = 0\right)$ for the toboggan (writing out expressions or symbols for each individual force).

d. What is the normal force of the ground on the toboggan?

e. What is the frictional force between the ground and the toboggan?

f. What tension does the rope need in order to keep the toboggan in equilibrium, so that its acceleration is 0? *Hint:* Solve Newton's second law for T. This tension is the force you would need to exert continuously while dragging the toboggan in order not to be pulled down the hill by it.

g. Let's say you finally make it to the top of the hill and jump in the toboggan, thereby making the tension 0 and giving the sum of the forces in the *x* direction a nonzero acceleration, resulting in you sliding down the hill. If the hill is 50 m tall and your mass is 70 kg, what will the speed of the toboggan be at the bottom? *Hint:* Use Newton's second law and the definition of acceleration. *Second Hint:* Do not try this at home. The hill is quite unreasonably steep. There are lots of trees in Minnesota, and toboggans are just about impossible to steer.

I. Having become slightly frozen on your last Minnesota winter vacation, you decide to vacation there during the summer. You are driving a pickup truck at a steady speed of 50 m/s up a hill with a slope 30° and coefficient of friction μ_k = 0.85, and, like any good Minnesotan, you are towing a boat with mass 3000 kg up to Lake Wobegon for a nice, relaxing summer day of ice fishing. *Hint*: If you are driving at a steady speed, the acceleration of the boat is 0, and therefore the boat is in equilibrium.
   1. Name the forces on the boat and state whether each force is acting parallel to or perpendicular to the slope.

   2. Draw a force diagram of the forces on the boat, including coordinate axes for the *x* and *y* directions.

3. By each set of forces, put a "+" or "−" sign to indicate whether the forces will be positive (added) or negative (subtracted).

4. Newton's second law will give you two equations $\left(\sum F_x = 0 \text{ and } \sum F_y = 0\right)$ and two unknowns. Which two forces are the unknown variables that we will be solving for?

5. Write down the first equation for this problem, $\sum F_x = 0$.

6. Write down the second equation for this problem, $\sum F_y = 0$.

7. Solve $\sum F_x = 0$ for the unknown variable that it contains which we are looking for.

8. Solve $\sum F_y = 0$ for the unknown variable that it contains which we are looking for.

9. State the magnitude of each of the four forces acting on the boat.

J. A 60-kg woman standing on a scale in an elevator is disconcerted to see that she's gained weight—the scale is now reading 700 N.
   1. What is the acceleration of the elevator?

   2. Need she be concerned about her weight, and why?

K. Your dear friend Susie has a mass of 200 kg, and she is getting concerned about her weight. Being in desperate straits, she turns to the local physicist—you—for help. You vaguely remember something from that lecture you half-slept through about elevators and weight, and you suggest that she try measuring herself in an elevator. Assume that the elevator (together with the scale) has a mass of 500 kg and is being held up by a cable. The elevator has an acceleration of ± 20 m/s^2, and that due to faulty design on the engineer's part it changes abruptly from positive to negative acceleration when it stops speeding up and starts slowing down. The scale reads in newtons. *Hint:* Normal force is always positive, and a scale will read 0 when nothing is touching it.

1. What does the scale read as it is accelerating upward?

2. What is the tension in the cable as it is accelerating upward?

3. What does the scale read as it is slowing down as the elevator reaches the top?

4. What is the tension in the cable as it is slowing down as the elevator reaches the top?

5. What does the scale read when the elevator is at rest at the top of the building?

6. What is the tension in the cable when the elevator is at rest at the top of the building?

7. Regrettably, the cable could not take the weight, and it snapped. Now the elevator is in free fall back toward the ground. Not one to waste a moment for physics, in the last few seconds before imminent death, you check the scale again. What does the scale say that Susie weighs now?

L. A big lead brick weighs 25 lb on earth.
   1. What is its mass in kilograms?

   2. What is its weight on earth in newtons?

3. The planet Torgon has a mass 5.8 times and a radius 0.3 times the mass of the planet Earth. What is the brick's weight on Torgon?

4. The moon has a gravitational acceleration 1/6 of g. What is the brick's weight on the moon?

5. Jupiter has a gravitational acceleration 2.64 times g. What is the brick's weight on Jupiter?

M. A freight train has a mass of $4.78 * 10^7$ kg. The locomotive provides a force of $1.3 * 10^6$ N, and the brakes provide a force of $-3.2 * 10^3$ N.
   1. How long will it take for the train to accelerate to its maximum speed of 15 m/s?

   2. The train accidentally stumbles into an action thriller movie, where the bad guys have blown up the bridge 750 m ahead. Will the train be able to stop in time?

N. A care package with mass 20 kg is dropped from an airplane traveling at speed $3.4 * 10^4$ m/s at a height 10 km onto a rural Third World village. It's a windy day, and the airplane is flying into the wind which exerts a force 25 N on the package.
  1. What is the magnitude of the package's acceleration?

  2. How long will it take the package to hit the ground?

  3. Where will it land relative to the position of the airplane when it was dropped?

O. A care package with mass m is dropped from an airplane traveling at speed v at a height h onto a rural Third World village. It's a windy day, and the airplane is flying into the wind, which exerts a force W on the package.
   1. What is the magnitude of the package's acceleration? Answer in terms of m, v, h, W, and g where $g = 9.8$ m/s^2.

   2. How long will it take the package to hit the ground?

   3. Where will it land relative to the position of the airplane when it was dropped?

P. Suppose the following 3500-N device is used as a horizontal ski lift to transport skiers between hills. Each side of the rope is tilted upward from the horizontal at an angle 10°.

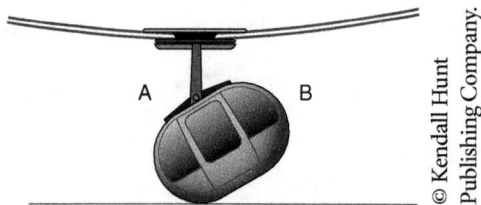

1. Find the tension in each cable.

2. Suppose cable A were raised higher. Would the tension in cable B increase, decrease, or stay the same?

Q. A traffic light is hanging from two cables tilted at 37° and 42°. If it has a mass of 2.3 kg, find the tension in each cable.

R. Suppose a 160-lb daredevil is hanging from an airplane that is accelerating upward with a magnitude 27 m/s² at an angle 72°. Not being the brightest bulb in the daredevil's club, he forgot to get his daredevil's permit, and in order to get away with his stunt, he had to sneak under the airplane and grab the only thing he could find—a loose electric cable that came unplugged and is hanging from the fuselage of the airplane. (The engineers approving the flight aren't the brightest bulbs in the aerospace industry either.)
  1. What will be the tension in the electric cable he is holding onto?

  2. Assume that it's a pretty tough cable, designed for a tough airplane, and it can withstand a force of 30,000 N, and the rope caught around his ankle is indefinitely tough. Regretfully, the plane is on a practice run for future astronauts learning to cope with strong accelerations. What is the maximum acceleration the airplane can have—at a 72° angle—before our dim-witted but courageous daredevil goes plummeting down to the ground?

Work and Energy   6

The physics we've done so far has been interesting, but not always as down to earth as it could *possibly* be. We'd like to describe a measurement of how hard it is to move something, and we were headed in the right direction with force when we defined inertial mass as resistance to acceleration. But force really *isn't* a measurement of how hard it is to move something. Anyone (perhaps with a little bit of weight training) can lift a suitcase full of lead bricks up to a height of about 1 m; that requires a pretty large force to counteract the gravitational force, but our muscles can cope with high demands for short periods of time. It takes a true athlete to take just two of those bricks and walk a hundred miles with them, even though the *force* applied is much smaller, since the gravitational force your force is in equilibrium with is smaller. It seems that a true measurement of "how hard it is to move something" depends not only on the force applied, but also the distance *over which* that force is applied. For this reason, a much better descriptor of how hard it is to move something is given by a simple concept called *work*, defined as follows:

$$W \equiv \boldsymbol{F} * \Delta \boldsymbol{x} \qquad (6\text{-}1)$$

This is a *dot product*, so work is a scalar—rather conveniently for us, since scalars are easier to work with than vectors (one doesn't have to take components of anything to add them or do anything else with them). One can calculate the dot product three ways. The easiest is by multiplying just the magnitudes together along with the cosine of the angle **between** the vectors, or $W = F\Delta x \cos \theta$. When the force is applied in the direction of motion (certainly the most efficient thing to do), $\cos 0° = 1$ and $W = F\Delta x$. When the force is applied perpendicular to the direction of motion, $\cos 90° = 0$ and one is wasting their effort. The second way to find the dot product is to multiply the x- and y-components of the force and displacement separately and then add them together; the third way (easier for those more mathematically inclined) is to project the force along the displacement, which in effect means taking the direction of the displacement as one axis, only keeping the component of the force along that axis.

Note that the definition of work given assumes that the force is constant over that displacement. This is usually the case—acceleration is usually constant, so force usually is too—but in cases where it isn't (such as a varying electric field) one has to integrate the force over dx, something only possible if the variation in the force is an integrable function of x. This book will avoid problems of that nature and take for granted the simplifying assumption of a uniform force.

We know from the previous chapter that $\mathbf{F} = m\mathbf{a}$, so let's play around with the definition of work by substituting m**a** for **F**:

$$W = m\boldsymbol{a} * \Delta \boldsymbol{x}$$

To keep things simple, let's assume one-dimensional motion, in which case

$$W = ma\Delta x$$

Now, to be clever, let's solve the fourth kinematics equation for a and substitute into the expression for work:

$$v^2 = v_0^2 + 2a\Delta x$$

$$a = \frac{v^2 - v_0^2}{2\Delta x}$$

So

$$W = \frac{m(v^2 - v_0^2)\Delta x}{2\Delta x}$$

Canceling out the displacements (ironically but quite fruitfully giving us an equation for work done **over a given displacement** but in terms of something other than that displacement),

$$W = \frac{m(v^2 - v_0^2)}{2} \tag{6-2}$$

$$W = \frac{1}{2}mv^2 - \frac{1}{2}mv_0^2$$

The expression $\frac{1}{2}mv^2$, for either initial or final velocity, is called **kinetic energy**, and consequently this way of writing the work is called the **work-kinetic energy theorem.** The v^2 comes from the fourth kinematics equation and just like before, it is really **v** * **v**; although we took a shortcut and assumed one-dimensional motion in our derivation, it turns out that this equation works for *any* motion, something slightly more complicated to prove. Kinetic energy is often abbreviated KE, and this theorem written as

$$W = \Delta KE \tag{6-3}$$

Both the definition of work given and the kinematics equation used to derive the work-kinetic energy theorem involve *displacement* rather than path length. While the work-kinetic energy theorem is true for work done by all forces, the work done by some forces (e.g., weight, or a spring) is independent of the path length, while the work done by others (e.g., friction, directly caused by the surface, if one may permit a medieval realist understanding of causality) *does* depend on the path taken. This doesn't violate our definition of work; any force that depends on the path taken is going to have a varying force, because unless the path taken is a straight line the force is going to have to change directions. When the force varies, one has to find the work by integrating over the path length. This is called a path integral, something also avoided in this book although a very useful skill to have under one's fingertips for quantum field theory, which requires doing an infinite number of them to solve problems, and the net effect is that the deviations from the shortest path cancel out, the same as if one had gone in a straight line. (The method by which one finds the shortest path, which in Euclidean geometry is a straight line, by canceling out all the deviations is called the calculus of variations, and was invented by Sir Isaac Newton overnight in order to solve a problem which someone had posed as a challenge, and is the result of what is possibly the world's most fruitful ego trip.)

A force which is independent of the path taken is called a "conservative" force (the reason being, as will be seen shortly, that when only these forces are present, an as-yet undefined quantity called "total mechanical energy" is conserved). Likewise, any force that *does* depend on the path is quite logically called a "nonconservative" force. Dividing forces into conservative and everything-but-conservative forces pretty much exhausts the forces one is likely to find. An old logical principle called the *principle of excluded middle* tells us that something must be or not be, with no third possibility—a force must be conservative, or non-conservative, but you won't find any forces which are both not conservative and not non-conservative at the same time. This principle, formulated by Aristotle, has been challenged by some modern logicians, and it had been challenged in ancient times in a logical treatise called the *Mulamadhyamakakarikas* by the Indian Buddhist writer Nagarjuna, and in modern physics it breaks down entirely when the full force of the implications of the postulates of quantum mechanics are brought to light. Nonetheless, it works good enough for us in classical physics. We can write the division between conservative and non-conservative forces mathematically as:

$$W = W_c + W_{nc} \tag{6-4}$$

where, not surprisingly, "W_c" stands for the work done by conservative forces and "W_{nc}" stands for the work done by nonconservative forces.

Like any other force, $W_c = F * \Delta x = F * x_f - F * x_0$. We can take a shortcut in our derivation by assuming that the force is constant along the path taken (which is not *ever* true of nonconservative forces), allowing us to express the work in this way without performing any integration, and then *define* **F** * **x** (for any *x*) as the negative of "potential energy" or $-U$. So by definition

$$W_c \equiv -\Delta U \tag{6-5}$$

What exactly is potential energy? For every conservative force independent of the path taken, the work needed to move an object using that force between two points in space is going to be the same regardless as to what path it took. The amount of work needed can therefore be thought of as a property of space and of the object being acted on in that space, rather than of the object "causing" the force, or even more precisely as a modification of space associated with the presence of a conservative force. (Again, we're being philosophically careful here. For one, the precise definition given avoids the philosophically slippery concept of "causality," a debate we'd like to bow out of whenever possible. Also, we haven't defined that ever-elusive concept of "space," only stated that at every coordinate point we can measure a quantitative scalar value for a potential energy. The property of space itself is called a "field," and mathematically and physically it's perfectly feasible to discuss vector fields as well where the potentials are vectors rather than scalars—we'll do this with magnetic forces. Conservative forces always yield scalar potentials, however. The potential energy in most cases—gravitational and electric potential energy, for example—is the field multiplied by some property of the object being acted upon, such as the mass or electric charge for the two examples respectively. This will become important in Chapter 10 when we distinguish "electric potential" or the field itself from "electric potential energy"; however, the distinction is not usually important for gravitational potential energies, or for nonelectrodynamic classical mechanics in general.)

Because this potential field ΔU is independent of the path taken, it can't be defined for nonconservative forces where **F** is or could be always changing direction. If an object is causing a force, then for any distance of space between two points there is a "potential energy difference" ΔU stating how much work will be done to move another object under the influence of that force between the two points. Obviously if the work were dependent on the path taken to traverse that distance, a potential difference between those two points in space could not be defined since there is always an infinite number of paths that could be taken to get between two points.

Let's substitute ΔKE for W in the equation above, and substitute $-\Delta U$ for W_c. Doing so gives us

$$\Delta KE = -\Delta U + W_{nc}$$

Rearranging,

$$W_{nc} = \Delta KE + \Delta U \tag{6-6}$$

This equation is called the law of conservation of mechanical energy. $\Delta KE + \Delta U$ together is called the change in "total mechanical energy" (that's why we defined the potential energy as *negative* the work done by conservative forces—so they would end up added together to get "total mechanical energy"), and the change in total mechanical energy is 0 when there are no nonconservative forces (and hence no W_{nc}). That's why these forces were called "conservative" forces, because they conserve energy within a system. If the system is the entire universe, since there is nothing "outside" the universe to perform a nonconservative force on it, the total mechanical energy remains constant. This tells us that mechanical energy, while still abstract and intangible, is somehow like a "thing." It says the same and doesn't change quantity. It only changes form. In fact, it will become significantly further reified when Einstein will discover that matter is like a form of potential energy, given by the famous equation $E = \gamma mc^2$ where $\gamma \approx 1$ for nonrelativistic motion. In special relativity, instead of speaking of "conservation of mechanical energy," we speak of "conservation of mass-energy". Nuclear reactions actually violate the classical law of conservation of mechanical energy if we do not understand mass as being like a form of potential energy).

Our argument so far has only proven that the *total* mechanical energy (or, relativistically speaking, mass energy) in the universe is constant. If this is true, then the creation or destruction of *any* energy requires the creation or destruction of the same amount of energy somewhere else in the universe. In other words, when a nonconservative force does work, thereby changing the mechanical energy of a system, that energy has to reappear somewhere else. Since we are not in quantum mechanics yet and would like to think of continuity in position for as long as possible (rather than what Einstein called "spooky action at a distance"), we think of the work done by nonconservative forces as dissipating energy from a system—actually taking energy present in a system and changing its form to transfer it out of the system. This mental image corresponds quite well with what we experience in everyday life. Friction is a nonconservative force, and rubbing two sticks together is perhaps the oldest and slowest and least effective method of starting a fire—it works (for those who, unlike this author during his Boy Scout years, had the patience to see it through to the end) because the work done by the friction dissipates the energy in the sticks in the form of heat, eventually raising the temperature until they are hot enough to spontaneously combust. Other examples of energy dissipated by nonconservative forces would include the light coming from a lightbulb, in which (due to the resistance of the tungsten filament, analogous to friction in a way) energy is transferred out of the system in the form of electromagnetic radiation or light. The sun continuously turns its mass into electromagnetic radiation, causing it to shine. An electrical generator continuously releases energy by inducing an electric current, requiring it to need to be turned constantly by a coal-powered steam engine or another such device.

When only conservative forces are present, the change in the total mechanical energy is 0, and

$$\Delta KE + \Delta U = 0 \qquad (6\text{-}7)$$

Cognizant that the symbol "Δ" means "final value minus initial value," we can rewrite this as

$$KE_f - KE_i + U_f - U_i = 0$$

We can rewrite this in two ways. Rearranging the terms gives us

$$KE_f + U_f - KE_i - U_i = 0$$

or

$$E_f - E_i = 0$$

where "E" is total mechanical energy, thereby (retroactively) justifying our earlier claim that $\Delta KE + \Delta U$ means change in mechanical energy. We could also instead add the negative terms over to the other side of the equation, giving us

$$KE_i + U_i = KE_f + U_f \qquad (6\text{-}8)$$

This is the form in which the law of conservation of energy is useful for solving problems. We can solve many of the same problems that we had in kinematics, but without having to deal with vectors at all, and only having one equation instead of four. (The only thing this doesn't tell us anything about is time—it was derived from the fourth kinematics equation, the only one of the four which doesn't have time in it.)

Kinetic energy is always $\frac{1}{2}mv^2$. But the equation for potential energy depends on which force is causing it. Until we encounter electric forces, we will only encounter two forms of potential energy: potential energy from the force of gravity and potential energy from the force of a spring. Knowing the equation for the potential we are dealing with, we can substitute these expressions for U in the law of conservation of energy and substitute $\frac{1}{2}mv^2$ for kinetic energy and solve for whatever variable is unknown; this method will work for pretty much any equation which does not ask for time and asks for height (or displacement) or velocity. It will sometimes work as a method for solving for mass, but only if the potential energy does not have mass as one of the terms (which is *not* true for gravitational potential energy). If nonconservative forces are present, *because they dissipate energy*, we can include that energy as the amount of heat dissipated, writing $E_f - E_i = Q$ where Q is the traditional symbol for heat.

Gravitational potential energy is easy to derive. Change in potential energy is defined as **F*Δx**, where the magnitude of **Δx** is always (for motion near the surface of the earth) going to be the height fallen *h*. (Note that we've already taken the liberty of projecting the displacement onto the force—mathematically equivalent to and in this case much easier than projecting the force onto the displacement—only worrying about the displacement in the *y* direction, the only direction in which gravity acts.) Gravitational force is the same as weight, which we already know to be *mg*, so the potential energy from the gravitational force is simply *mgh*.

For example, if a ball is dropped from rest from a height 72 m and we are asked to find the final velocity when it hits the ground, instead of using a kinematics equation it is very simple to use the law of conservation of energy. There are no nonconservative forces involved, just gravity, and no other sources of potential energy mentioned.

$$KE_i + U_i = KE_f + U_f$$

The initial kinetic energy is 0 because it is dropped at rest, and the final potential energy is 0 because we typically define the ground level to be $h = 0$. (The height *h* is just a convenient term for position along the *y*-axis; the coordinate we assign it is arbitrary, and only the *difference* in *h* or displacement has physical meaning.) Therefore in this case

$$U_i = KE_f$$

or

$$m(9.8)(72) = \tfrac{1}{2}mv^2$$

The masses cancel (as they should—we can solve this using kinematics, which doesn't employ the concept of mass at all), giving us

$$(9.8)(72) = (0.5)v^2$$
$$v^2 = 9.8(72)(2) = 1.41 * 10^3$$
$$v = 37.6 \text{ m/s}$$

One should hopefully recognize 9.8(72)(2) as being $2a\Delta x$ from the fourth kinematics equation, with $v_0^2 = 0$ because it was dropped from rest. Conservation of energy is usually nothing more than a convenient way of writing the fourth kinematics equation.

Example 6.1

Dropping a ball from a certain height is something that could have been done with the fourth kinematics equation, it's a simple one-dimensional problem. What is powerful about the fourth kinematics equation, which conservation of energy will capitalize on, is that each of the terms in the equation $v^2 = v_0^2 + 2a\Delta x$ is a *scalar*. All of them are dot products between vectors, and can be written $v * v = v_0 * v_0 + 2a\,\Delta x$, where we don't specify that each of the vectors v, v_0, a, or Δx is the component of the original vector projected along some axis. We can actually take the entire magnitude of v, and the entire magnitude of v_0, and since the dot product used to square each of them gives us a scalar that is just a regular number that can be added in a normal fashion, when the fourth kinematics equation or conservation of energy is used, it doesn't matter whether v and v_0 are pointing the same direction. The only requirement is that a and Δx point the same direction, or at least that the angle between them is known so that the dot product can be calculated as $a\Delta x \cos \theta$.

In question E from the homework, an example is taken of a cart sitting on top of a hill. The hill has a rugged, uneven shape—something that would have been a problem in Chapters 4 and 5, since the normal force would constantly be changing as the angle of the slope changes, and therefore the sum of all the forces and the cart's overall acceleration would be changing, and everything we did in Chapters 3, 4, and 5 presumed a constant acceleration. Because we have normal force acting on the cart, this isn't a free-projectile problem like the ones we saw in Chapter 4.

However, taking a y-axis going straight up and down in the vertical direction in the picture below, and taking the x-axis going in the horizontal direction in the picture, we can see that the y-component of the acceleration will be constant—it's just -9.8 m/s². And we know the y-component of the displacement—it's the height of the hill, h, given in problem E as 23 m. At any given point on the hill, v and v_0 will be pointing different directions. But that's okay—since the fourth kinematics equation is a relationship between scalars, we can still use it—and therefore use conservation of energy—to solve the problem. In the homework, the student will answer some qualitative questions about this problem; right now, let's find the cart's speed at the bottom of the hill. The cart is show n in the picture with wheels; so far we've assumed that objects have no *internal* motion, including spinning wheels, since they carry their own "rotational kinetic energy," which we can calculate in Chapter 8, but we can simplify things by continuing to ignore that for now.

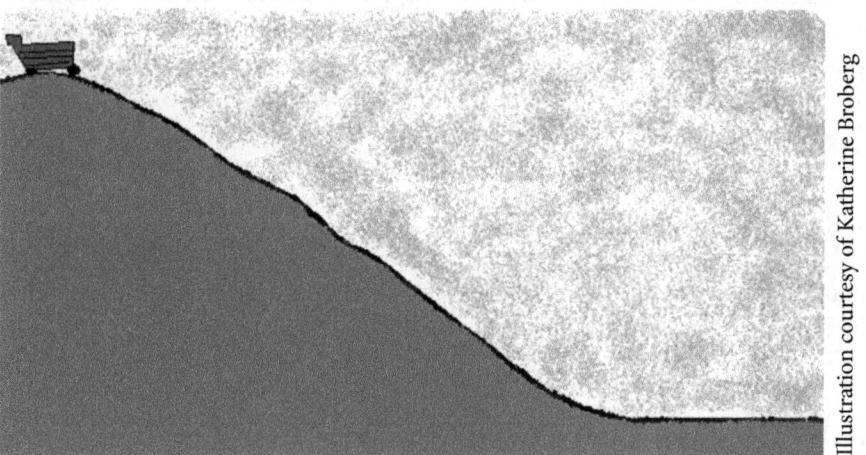

Illustration courtesy of Katherine Broberg

The cart is at rest at the top of the hill, and it is on the ground ($h = 0$) at the bottom, so conservation of mechanical energy tells us that

$$\frac{1}{2}mv^2 = mgh \tag{6-9}$$

The masses cancel out—as they always will, when the only force is the gravitational force—and we can simply plug in the value 9.8 for g and the value 23 m for h and solve for v. Solving for v,

$$v = \sqrt{2gh} \tag{6-10}$$

giving us $v = \sqrt{2(9.8)(23)} = 21.2$ m/s.

Using *mgh* as gravitational potential energy only works near the surface of the earth. Problems involving motion in outer space, where the gravitational acceleration *changes* over long distances, are more tricky. For these, we have to return to Newton's universal law of gravitation to find a more general definition of gravitational potential energy:

$$F_g = G\frac{M_1 m_2}{r^2}\hat{r}$$

Since $U = -W = -\mathbf{F} * \mathbf{r}$ where **r** is displacement in polar coordinates,

$$U = -G\frac{M_1 m_2}{r^2}\hat{r} * \mathbf{r} = -G\frac{M_1 m_2}{r} \tag{6-11}$$

Note that *by convention* we can change the reference point $U = 0$ to be anywhere we want, which is mathematically acceptable because this expression for U can also be derived with greater generality by taking the *integral* of $\mathbf{F} * d\mathbf{r}$ over all **r**, and any integral has an undetermined integration constant we can add to the equation. Using the definitions we started with (which imply the symmetries that allow us to avoid calculus), we get one possible expression of the potential, namely that with integration constant 0. Doing so shows that U will approach 0 as r approaches infinity, sometimes (in the case where we are calculating escape velocity) but not always the most useful reference point. The point to take away is that since it is only the *difference* in potential energy that has physical meaning, which is to say that since this is an integral we can add *any* scalar constant of integration without changing the actual physics, that we can still choose any point we want such as the ground, for the potential energy to be 0.

It is useful to set potential energy equal to 0 at infinity when calculating escape velocity from a planet, however. Escape velocity is defined as the velocity needed to barely but *completely* escape the planet's gravitational field. Of course, Newton's universal law of gravitation has no "distance limit" on it (*r* can be arbitrarily large, making the gravitational force arbitrarily small without actually becoming 0), and in order to *completely* escape the planet's gravitational field it needs to be able to get infinitely far away from it, which, of course is only an abstraction. The escape velocity is when a rocket or other vehicle *barely* escapes—the minimum velocity needed to get away without falling back down to earth—so we state that it is the initial velocity for which the projectile's velocity does not slow down to 0 until it is infinitely far away. (If it is 0 any closer, it will become negative just beyond that point, and will fall back down to earth.)

The gravitational potential energy expressed this way is negative because potential energy *increases* the farther you get away from the source of the force—it takes a positive amount of work to lift an object away from the earth. Since it takes a positive amount of work to lift an object away from the earth, the principle of inertia tells us that the gravitational force one is acting against is *negative*—simple enough, since it is pointing downward. The work done *by* gravity is therefore negative, and since the change in potential energy is *equal and opposite* to the work done *by* the force, the change in potential energy must be positive, so the potential at a higher position must be higher than the potential at a lower position. But if the potential is 0 at infinity, and this must be *higher* than the potential at the surface of the earth, the potential at the surface of the earth must be *negative* or $-G\frac{M_1 m_2}{r}$ rather than $+G\frac{M_1 m_2}{r}$; QED.

Plugging these values into the law of conservation of energy, we find that

$$\tfrac{1}{2}mv_{esc}^2 - G\frac{M_1 m_2}{r} = 0$$

where the terms on the left side are the initial kinetic and potential energy and the 0 on the left side is the final total mechanical energy. Solving for the escape velocity v_{esc},

$$\tfrac{1}{2}mv_{esc}^2 = G\frac{M_1 m_2}{r}$$

$$\tfrac{1}{2}v_{esc}^2 = -G\frac{M_1}{r}$$

$$v_{esc}^2 = 2G\frac{M_1}{r}$$

$$v_{esc} = \sqrt{2G\frac{M_1}{r}} = \sqrt{2gr} \qquad (6\text{-}12)$$

This holds true wherever $g \equiv G\frac{M_1}{r^2}$ is the gravitational acceleration from one's initial position and r is the distance from one's initial position to the center of mass of the planet. Note that if one had not noticed that gravitational potential increases with distance, we would have gotten an escape velocity $\sqrt{-2gr}$ which is an imaginary number and physically meaningless to apply to a velocity—there is no escape velocity because one would have already escaped from the planet. Potential will *always* be higher the farther away the two objects are, with one exception: the repulsive electric force between two charges *of the same charge* in which case the potential will decrease (because it takes work to push them together rather than pull them apart).

The second potential that is convenient to discuss at the present moment is the potential from a spring. Springs haven't been discussed yet. Springs exhibit oscillatory motion, so the force is constantly changing both magnitude and direction, causing the spring to bounce up and down rather than accelerate uniformly. Fortunately, there is an easy (and completely empirical) way of describing this force without trying to derive it from first principles and hideous calculus. This empirically determined fact is called Hooke's law:

$$F = -kx \qquad (6\text{-}13)$$

Here x is the conventional way for writing the displacement from the equilibrium position where the force is 0. It is the same as Δx where x_0 is that position such that the force is 0—the spring is neither stretched nor compressed. Hooke's law has a negative sign because the spring gives a restoring force—when the spring is stretched it accelerates back until it has reached its equilibrium position. The displacement is positive, so the force must be negative until it reaches the equilibrium position with F = 0. Because it has been given a speed by this force, the spring will continue to compress even though the force is 0 (remember, force is proportional to *acceleration* not velocity), but the force continues increasing past 0 and becomes positive, causing the speed to slow down until it reaches its maximum compression (in which x will be a negative number, still being the displacement from the equilibrium position) at which point its speed will be 0, with a maximum force being applied to it. It will, of course, bounce back to its equilibrium position and keep going until it reaches the position it was initially stretched to, bouncing back and forth indefinitely unless friction slows it down.

The constant k is called the "spring constant" and must be measured experimentally for each spring.

Since potential energy is defined as $F * x$, making the change of terminology "x" for "Δx" now that we are talking about springs, it is a simple matter to find the potential energy of the spring. It is only complicated by the fact that F is not constant. It takes work to stretch a spring from its equilibrium position out to some distance d; to find the potential energy rigorously, we must integrate $\int_0^d F(x)dx$. We're fortunate in that Hooke's law is a linear equation, and if we graph $F(x)$ versus x we get a straight line starting at the origin and ending at d on the x-axis. The integral is the area under the curve which forms a triangle, and the area under a triangle is $\frac{1}{2}bh$ from basic geometry. The length of the base is d and the height is the force at $x = d$, which is given to us from Hooke's law as $-kd$. So the area under the curve, or the potential energy, is $\frac{1}{2}kd^2$. In general, for any distance x from the equilibrium position, the potential energy is $\frac{1}{2}kx^2$, the form in which it is usually written.

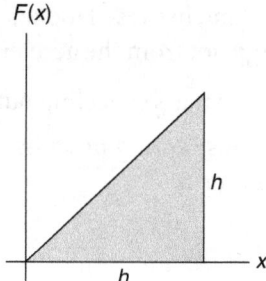

This expression can also be found by averaging the initial and final forces, with the initial force (at the equilibrium position) being 0. This is mathematically acceptable in this case because the force increases uniformly.

Let's do a quick example of a problem involving a spring and spring potential energy.

Example 6-1 Suppose that a 20-kg block is compressing a spring by 2 m on a horizontal, frictionless surface. If the spring constant is 700 N/m, how far will the block go up a frictionless incline with an angle 35°?

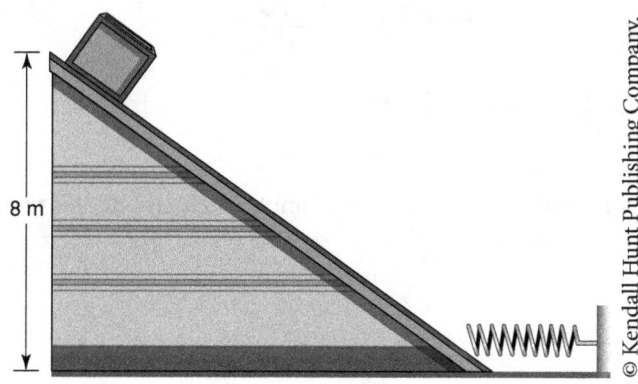

Now the potential energy includes *both* gravitational and spring potential energy, and both must be included in the law of conservation of mechanical energy.

$$KE_i + U_i = KE_f + U_f \qquad (6\text{-}14)$$

$$\frac{1}{2}mv_i^2 + mgh_i + \frac{1}{2}k_i x^2 = \frac{1}{2}mv_f^2 + mgh_f + \frac{1}{2}k_f x^2$$

The initial kinetic energy will be zero (it starts at rest) as will the initial gravitational potential energy (defining h to be ground level as usual). The final kinetic energy will be 0 if we want to know how high it goes (at what height is the peak, where the y-component of the velocity is always 0; the normal force of the ramp will force the x-component of the final velocity to be 0 as well), and at its peak the block will not be touching a spring so k_f will be 0.

That leaves us with

$$\frac{1}{2}k_i x^2 = mgh_f$$

Plugging in known values,

$$(0.5)(700)(2^2) = (20)(9.8)\, h_f$$

Note that the mass is now important, because Hooke's law is independent of the mass and we derived the spring potential energy not from the acceleration (which we would only be able to write as $-\frac{kx}{m}$ anyway, resulting in the mass canceling out in the spring potential energy) but rather directly from Hooke's law. Because spring potential energy is independent of the mass, the mass does not cancel out in every term.

Solving for h_f, we find

$$h_f = 7.14 \text{ m}$$

Note that the angle of the incline didn't go into our calculations. Because the incline is frictionless, we don't care how far it went in the x direction; we just care how high it went. A problem like this might ask how far along the ramp it went, and in that case the angle is important (since the distance along the incline is $\frac{h}{\cos \theta}$, from the geometric definition of the trigonometric relations).

Nothing that has been discussed so far has involved time. To bring time into our discussion of work and energy, we introduce one simple concept—the rate at which work is done, or power.

$$P = \frac{W}{t} \qquad (6\text{-}15)$$

Like work and energy, power is also a scalar. Since $W = \mathbf{F} * \mathbf{\Delta x}$,

$$P = \frac{\mathbf{F} * \mathbf{\Delta x}}{t}$$

Since $\frac{\Delta x}{t}$ is defined as average velocity,

$$P = \mathbf{F} * v_{ave} \qquad (6\text{-}16)$$

The force, of course will cause an acceleration; the velocity here is the *average* velocity of the object during a given *time* frame.

HOMEWORK FOR CHAPTER 6

Name _____

A. Standing at the edge of a cliff 240 m tall, you throw two balls into the air—one directly upward at 5 m/s and another directly downward at −5 m/s. Use conservation of energy to show that they will have the same final speed at the bottom. Show your work.

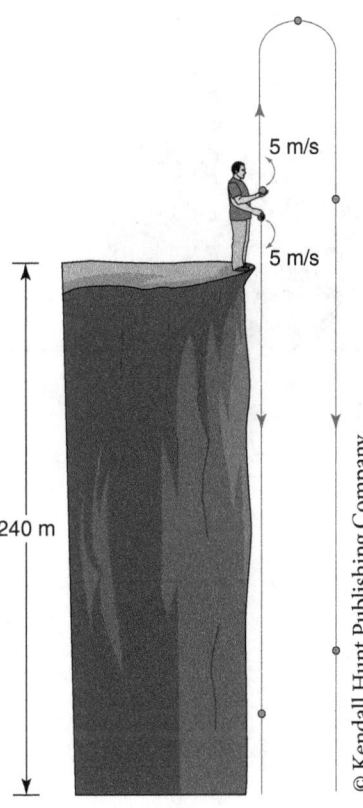

B. A cannon launches a cannonball with an initial speed 210 m/s at a 70° angle from the horizontal.

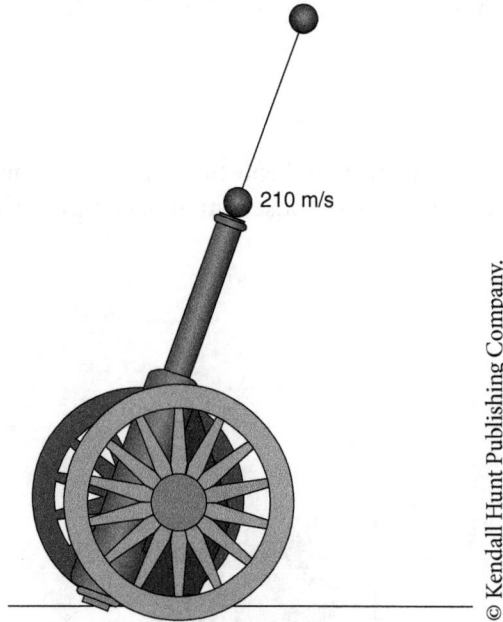

1. How high will the cannonball go?

2. If the cannonball has a mass of 25 kg, how much work does it take to launch the cannonball to that height?

C. A spring BB gun fires a BB at a speed of 35 m/s. If the barrel is 1.5 m long, and the mass of the BB 0.001 kg, what is the spring constant of the gun?

D. Suppose for some unexplained reason you were skateboarding (or skiing, if that's your taste) and wanted to use a spring to launch you off a ramp. Assuming you have a mass of 70 kg, if the spring has a spring constant $k = 625$ N/m and is compressed 5 m,

1. If the ramp has a 60° slope and is 10 m long, will you make it off the ramp?

2. How high will you go?

3. How fast will it be going when it enters the ramp? (Ignore friction.)

4. Suppose the ramp were not frictionless, and you only make it up 4 m high. Find the heat (in joules) lost due to friction.

5. Given your answer to question 4, find the frictional force and the coefficient of kinetic friction.

E. A cart with a mass of 25 kg starts at rest on top of a hill 23 m tall.

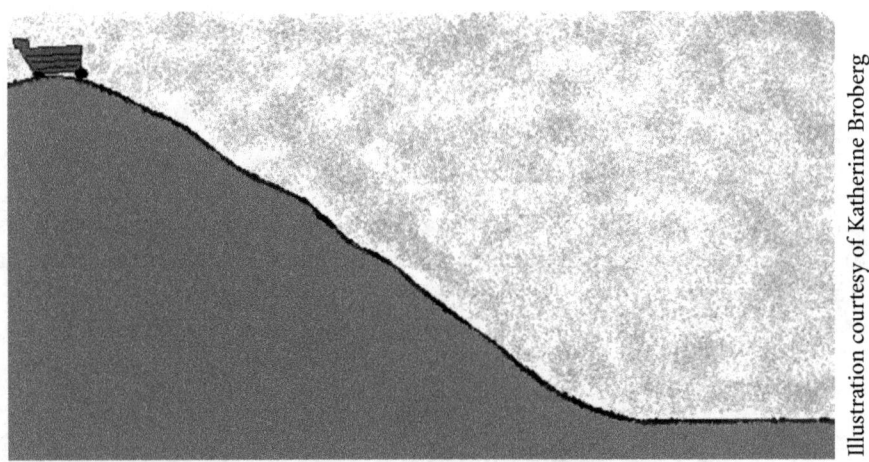

1. Where is the kinetic energy 0 for this story?
   a. Is the potential energy 0 here?

2. Where is the potential energy 0 for this story?
   a. Is the kinetic energy 0 here?

3. Is either the kinetic or potential energy 0 in the middle of the slope?

4. Show how the magnitudes of the kinetic, potential, and total energies compare to each other on the ground, at a height of 10 m, at a height of 20 m, and at 23 m. (e.g., TE > KE = PE, or TE > PE > KE)

5. What is the total change in mechanical energy?

F. A man is standing on top of a building 25 m tall throwing a ball with speed 5 m/s at the following angles: 0°, 45°, 80°, 90°.

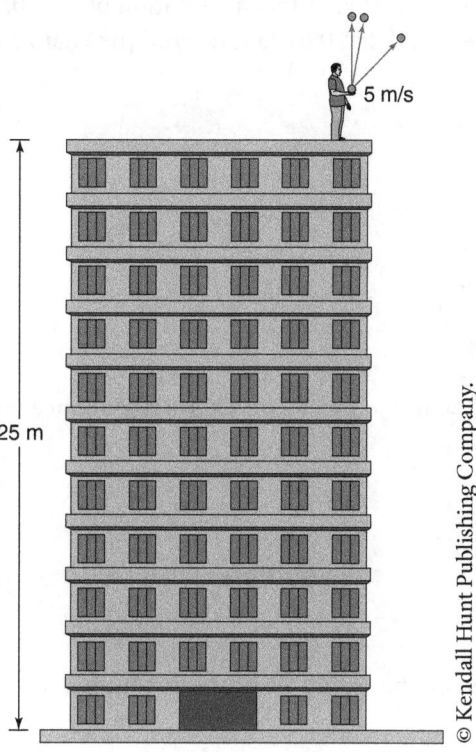

1. Does increasing the angle at which the ball is initially thrown make the magnitude of the final velocity of the ball less than, equal to, or greater than the magnitude of the final velocity of a ball thrown at a lower angle?

2. Does increasing the angle at which the ball is initially thrown make the magnitude of the final mechanical energy of the ball less than, equal to, or greater than the magnitude of the final mechanical energy of a ball thrown at a lower angle?

3. Does increasing the angle at which the ball is initially thrown make the total time the ball spends in the air less than, equal to, or greater than the total time spent in the air of a ball thrown at a lower angle?

G. Suppose at an amusement park you are sitting in a cart with mass 200 kg resting against a spring compressed by 5 m. The cart follows a frictionless track with a circular loop oriented vertically. The cart is launched with an initial velocity 25 m/s.
   1. The force of the spring acting on the cart lasts for the duration of the spring's contact with the cart. During this time, the cart accelerated from 0 to 25 m/s. Find the change in kinetic energy of the cart.
   Hint: $KE = \frac{1}{2}mv^2$

   2. Find the work done by the spring on the cart. *Hint:* This is trivial once you know the answer to question 1.

   3. Find the potential energy stored by the spring while it was compressed.

   4. Find the spring constant. *Hint:* $\Delta U = -\frac{1}{2}kx^2$

   5. Find the total potential energy at the bottom of the loop.

6. Find the total kinetic energy at the bottom of the loop after leaving the spring and immediately before climbing up the loop. *Hint:* This is trivial once you know the answer to question 1.

7. Find the total mechanical energy at the bottom of the loop. *Hint:* This is trivial once you know the answers to questions 5 and 6.

8. Use the law of conservation of mechanical energy to find the maximum radius the loop can have before the cart falls off. *Hints:* (a) The height the cart will go will be twice the radius of the loop. (b) There are no nonconservative forces, so the maximum height is where all of the spring potential energy that it had at the bottom will have been converted to gravitational potential energy.

9. Draw a force diagram of the cart at the top of the loop. Label all the forces clearly.

10. For circular motion, it so happens that Newton's second law can be expressed for the *radial* and *tangential* directions rather than for the *x* and *y* directions, which makes our life a lot easier. The radial direction points from the center of the loop toward the cart, regardless of where the cart is on the loop. For circular motion, the radial component of acceleration is called *centripetal acceleration*, and is equal to $a_r = \dfrac{v^2}{R}$ where R is the radius of the loop. v^2 is the dot product between the total velocity and itself, or **v** * **v**—so at the top of the loop v^2 will be the square of the magnitude of the velocity. If the radius is as large as it can possibly be without the cart falling off, the normal force of the cart will just approach 0. Set the normal force equal to 0 and use Newton's second law to find the cart's velocity at the top of the loop.

H. Suppose that you are on the planet Zilborg, which has an average radius $1 * 10^9$ m and average density 7.4 kg/m³.
   1. Assuming that Zilborg is a perfect sphere, find the mass of the planet.

   2. Using Newton's universal law of gravitation, find the gravitational acceleration on the surface of the planet.

   3. The escape velocity on Zilborg is the *minimum* speed the rocket needs in order to never fall back down to the surface again in a universe in which Zilborg is the only object. (In real life, the escape velocity can be a little bit lower because Zilborg's gravitational force will eventually be overpowered by the gravitational force of other stars and planets. In real life, there is no planet Zilborg. I made it up. They have pills for people who think they are on planet Zilborg.) In order for the escape velocity to be a speed at which it can escape Zilborg's gravitational pull, it must be able to travel an infinite distance away from Zilborg without falling back down first. In order for the escape velocity to be the *minimum* speed at which the rocket can completely escape Zilborg's gravitational pull, what will the rocket's velocity be when it finally reaches a position an infinite distance away? In the next problem, this position will be called x_f.

4. What is the kinetic energy at x_f? *Hint:* This is an easy question.

5. Using Newton's universal law of gravitation, the definition of work as W ≡ **F** * **Δx**, and the fact that work done by a conservative force is the change in potential energy, what is the change in potential energy of the rocket? The rocket has a mass of 700 kg.

6. What is the potential energy at x_f?

7. What is the potential energy of a 700 kg rocket at x_0, the surface of Zilborg?

8. Using the law of conservation of energy, find the initial kinetic energy of the rocket.

9. From the initial kinetic energy of the rocket, find the rocket's initial velocity. This is the escape velocity of the rocket.

I. Having successfully escaped the planet Zilborg, unfortunately your navigator forgot to check for planets in the way, and 34 million light years away you end up crashing headlong into the asteroid Broborg. (Do not ask me why an asteroid would be floating 34 million light years into the middle of nowhere. Astrophysically speaking, it's kind of improbable. But some of these asteroids are strange fellows!)

1. How far away from the center of the planet Zilborg is the asteroid Broborg in meters? *Hint:* A light year is the distance light crosses in a year in a vacuum. Light travels at $3.00 * 10^8$ m/s.

2. The rocket has a mass 700 kg (and for the sake of simplicity, it uses a high-tech antimatter engine with a limited fuel supply, meaning that after a very short time it becomes a free projectile and there is no measurable change in the rocket's mass). Find the gravitational potential energy between the rocket and Zilborg when the rocket is 1 light year away from crashing into Broborg.

3. Broborg is an irregularly shaped object with a pointy end and a total mass $1 * 10^{17}$ kg. (As you can see, it is rather dense.) The rocket will crash into Broborg at a point where the surface of the asteroid is 1 km from the center of mass. Find the gravitational potential energy between the rocket and Broborg when the rocket is 1 light year away from crashing into it.

4. Assuming that the gravitational potential energy between the rocket and Broborg is small enough to ignore when the rocket is 1 light year away from crashing into it, use the law of conservation of energy with the gravitational potential energy from Zilborg to find the rocket's velocity when it is 1 light year away from crashing into Broborg.

148 Foundations & Principles of Physics

5. As the rocket comes crashing into Broborg, there are two forces acting on it: the gravitational force from Broborg giving it a positive acceleration and the gravitational force from Zilborg giving it a negative acceleration. Newton's laws tell us that the total force on the rocket is the sum of the two gravitational forces, but unfortunately since the strengths of those forces change over distance we need calculus to solve the problem using Newton's laws. Fortunately, we can use conservation of energy to bypass this mathematical difficulty. Start by finding the gravitational potential energy between the rocket and Zilborg as the rocket is crashing into Broborg.

6. Find the gravitational potential energy between the rocket and Broborg as the rocket is crashing into Broborg. *Hint:* It's not 0, since for both the planet and the asteroid we have to define the 0 point as being at infinity.

7. Subtract the gravitational potential energy between the rocket and Broborg from the gravitational potential energy between the rocket and Zilborg in order to find the total gravitational potential energy relative to Zilborg's coordinate system, at the point when the rocket is crashing into Broborg.

8. Find the distance from the center of mass of Zilborg where the gravitational potential energies from Zilborg and Broborg cancel out so that the total gravitational potential energy is 0, with an object remaining in a stable equilibrium between the two bodies. (This is similar to and somewhat simplified from what is called a *Lagrange point* in classical mechanics, a stable point in an orbit between two bodies. In all five seasons of *Stargate Atlantis*, "Lagrange point" was the only scientific term used correctly, found season 1, episode 19, "The Siege, Part 1." A Lagrange point is actually the place where the combined gravitational force from two nearby objects—planets or stars—causes centripetal acceleration in an orbit around both objects.)

9. Find the change in total gravitational potential energy between the rocket's position 1 light year away from impact and its actual impact with Broborg.

10. Use the law of conservation of energy to find the rocket's velocity when it comes crashing into Broborg.

11. Recall that we defined momentum as $\mathbf{p} \equiv m\mathbf{v}$. Find the rocket's momentum just before it crashes into Broborg.

12. The asteroid Broborg was peacefully motionless and at rest before your rocket plummeted into it. After the rocket crashes into it, it becomes buried in the interior of the asteroid in a perfectly inelastic collision. Use the law of conservation of momentum to figure out the asteroid-*cum*-rocket's final velocity. *Hint:* Conservation of momentum says that $\mathbf{p}_{rocket} + \mathbf{p}_{asteroid} = \mathbf{p}_{mess\ after\ collision}$.

13. The law of conservation of energy tells us that the work done by nonconservative forces is the change in total mechanical energy of the system. Friction is a nonconservative force, and it takes a fair amount of friction to slow down a falling rocket. Assume that the rocket ends up buried only 10 m deep inside the asteroid, so that the difference in gravitational force from the surface to its final position is negligible. (Otherwise you have to use calculus.) Find the change in gravitational potential energy from the moment of impact to the point where the rocket is buried 10 m into the asteroid.

14. Using your answer to question 13, find the work done by friction to slow down the asteroid.

15. Using the definition of work, find the frictional force applied by the asteroid.

16. After a hard day of digging the rocket out (and what is a "day" on an asteroid that isn't revolving on its axis and isn't orbiting any star, anyway?), you unfortunately discover that the engines are hopelessly broken. The only fix is to head back to Zilborg where they have all the finest mechanics. But you need a way to get the engine off the planet. Fortunately, your rocket is equipped with a plasma cannon that you keep around in case of unfriendly aliens, and your engineer has a brilliant idea. One of you will walk to the other side of Broborg and fire a 25-kg plasma cannonball through the asteroid at your rocket. Fortunately, the asteroid Broborg, due to unusual circumstances, and, even more fortunately, the cannonball have the same charge, so it can plow through the asteroid without encountering any friction as all of the atoms in the asteroid will simply move to the side. Of course, the cannonball will lose energy as it does work pushing atoms out of its way, but your friendly volunteer brought a radar gun to track its velocity as it moves. The cannonball is released from the cannon with an initial velocity 700 m/s (we're talking heavy-duty alien smashing machines here), and, after making its way 2 km to the other side of the asteroid, your friendly volunteer's radar gun registers a velocity 250 m/s. (Unfortunately this brilliant plan forgot to include a way for the volunteer to get back on the spaceship—minor technicality.) Assume that the cannon and volunteer together have a mass 100 kg, and that together with the cannonball they have gotten off the rocket. The cannonball makes a perfectly *elastic* collision with the rocket, meaning that no heat is released—they just bounce off each other like billiard balls, since now the rocket has picked up an electric charge from the asteroid and there is no actual impact. The electrostatic repulsion from the cannonball makes the rocket go flying off toward Zilborg. Since no heat was released from the collision, the total mechanical energy of the system remains unchanged, and because no *external* force was applied on the system, the total momentum of the cannonball and the rocket together is conserved (just as with an inelastic collision). Use the laws of conservation of momentum and conservation of mechanical energy to solve for the velocities after the collision of the cannonball and rocket.

17. What is the heat released by the work done on the cannonball by the asteroid?

18. What is the escape velocity of the rocket on Broborg? *Hint:* It is the velocity needed to slow down to a velocity of 0 at the Lagrange point between Zilborg and Broborg. After it passes that point, it will be falling toward Zilborg and gaining speed.

19. Given the velocity calculated in question 19, will the rocket make it back to Zilborg, or will it fall back to Broborg leaving the crew stranded and helpless?

Conservation of Linear Momentum and Collisions 7

In this chapter, we're going to take a step back from the work we've done and re-derive Newton's Laws, using a formulation closer to the one Newton himself actually used. (There are actually quite a few different formulations in basic mechanics, and in upper-level mechanics classes most of the material we've covered so far will be recast in the Lagrangian and Hamiltonian formulations of physics, which are quite useful but involve too much calculus to worry about here.) The last problem in the last chapter hinted at the definition of a new composite concept, *momentum*.

Recall that we presented Newton's second law as an extension (using the principle of superposition, or Newton's first law) of the definition of inertial mass, which we gave as

$$\boldsymbol{F} = m\boldsymbol{a} \tag{7-1}$$

Now average acceleration is a basic kinematics concept, defined as the rate of change of velocity:

$$\boldsymbol{a} = \frac{\Delta \boldsymbol{v}}{t} \tag{7-2}$$

and force can consequently be written as

$$\boldsymbol{F} = m\frac{\Delta \boldsymbol{v}}{t}$$

In most cases we'll see, the mass of an object isn't going to change. We've kept things simple by ignoring the internal structure of objects in motion; we just care right now (and for most of the remainder of this text) about how the object as a whole moves and interacts with things. Consequently, we are assuming that in general the mass of the object or system is not going to change. There are cases in which the mass will change, of course—explosions, chemical changes, collisions resulting in objects breaking, and the occasional nuclear reaction that happens every now and then. But all of them—as well as Dr. Feynman's quirky "relativistic mass"—are going to be outside the scope of our survey. Even if we do a problem applying our physics to subatomic particles in which case masses are much more fluid and variable than they are in the macroscopic world (since collisions involve particles being destroyed and new and possibly different particles being created out of the energy), we'll consider the *mass-energy of the entire system*, which will be invariant – if any mass were to be destroyed or created in a nuclear reaction, the amount of mass is equal to an energy given by Einstein's famous formula, $E = \gamma mc^2$ where γ is a relativistic correction that is 1 at slow speeds. Problems in this book will avoid these kinds of complications.

So because the mass is going to be (for almost all purposes) invariant, we can employ the distributive property of algebra to rewrite the expression for force:

$$F = m\frac{\Delta v}{t}$$

$$F = m\frac{v_f - v_0}{t}$$

$$F = \frac{mv_f - mv_0}{t}$$

$$F = \frac{\Delta(mv)}{t} \tag{7-3}$$

The quantity mv is called **linear momentum.** The equation above applies only to an *average force* over a given time interval t, a generally reasonable simplification since forces and accelerations tend to be constant. As with our other quantities, we can if we need to extend this expression from average force to instantaneous force by taking the limit as $t \to 0$, during the "Δ" into a differential:

$$F = \frac{dp}{dt} \tag{7-4}$$

using Leibniz's nomenclature (usually employed in mathematics classes), or to use Newton's own system of denotation (more common in upper-level mechanics courses).

$$F = \dot{p}$$

This was how Newton originally conceived his second law. A force is the rate of change of momentum. Consequently, if no force is applied, momentum is a *conserved quantity*, a quantity which does not change, and we are given the law of conservation of linear momentum:

$$p_i = p_f \tag{7-5}$$

or

$$\sum_i m_i v_i = \sum_f m_f v_f \tag{7-6}$$

We've seen two conserved quantities now: mechanical energy and linear momentum. Let's review when and how these quantities are conserved, in order to avoid future confusion:

- *Mechanical energy* is conserved when no nonconservative forces are applied to the system
- *Linear momentum* is conserved when no *net external forces* are applied at all to the system

I've been careful to use the word *system* rather than *particle*. That's because a system can include a set of particles—in particular, when particles collide with each other, and it's much easier to treat all of the particles in collision as a single system (so that the conservation laws can be held true) rather than looking at individual particles and trying to calculate the forces acted on them.

One might easily but mistakenly think that conservation of mechanical energy is more generally applicable than conservation of linear momentum (or to use precise logical language, that the law has greater *extension*). After all, conservation of mechanical energy only restricts us to conservative forces, while conservation of linear momentum rules out the possibility of having any net forces being applied *at all*.

A closer look at the language will show that the opposite is the case, however. These two laws will come in handy mostly during collisions, where there certainly will *always* be forces involved—but if we treat all the particles colliding as a whole system, then the forces *between* those particles are internal forces and there is still no *external* force being applied on the system. In fact, all collisions—except contrived situations like charged particles colliding in a magnetic field, which we won't punish ourselves with—will obey conservation of linear momentum.

But not all collisions will obey conservation of mechanical energy. Conservation of energy requires that no nonconservative forces be applied to the system—without specifying whether those nonconservative forces are internal forces or external ones. And *any* time a collision results in the deformation of the objects colliding, whether it be the give and take of the surface of a rubber ball or the complete structural collapse of a vehicle in a road accident, is going to require work, work which will be released in the form of heat. That heat has to come from somewhere, and it comes from the kinetic energy of the particles. Work is being done by a nonconservative force, and even though this force is *internal*, heat is still radiated out of the system.

A collision in which both mechanical energy and linear momentum are conserved is called a *perfectly elastic* collision. The best model one can think of to envision them is the collision of hard billiard balls, which do not deform (at least not noticeably) upon impact and therefore do not lose any kinetic energy in the course of the collision. The only collisions in real life that come truly close to a perfectly elastic collision are collisions between subatomic particles, but perfect elasticity is a *fairly* accurate approximation when the raise in temperature due to the collision is negligible.

The other extreme is called a *perfectly inelastic* collision, in which the two objects colliding are so deformed and jumbled together that they become stuck together as one object having their combined mass, moving at the same velocity. Because they are moving at the same velocity, we need to solve only for one velocity, and such problems can be solved with only one equation instead of two. Complete car wrecks aside, most collisions are not perfectly inelastic—they lie somewhere in-between, and in order to solve problems asking for the final velocity one must know how much heat was released in the process, a typically difficult thing to measure in everyday situations. For this reason, most textbook problems limit themselves to perfectly elastic and perfectly inelastic collisions.

Let's give an example of a perfectly inelastic collision. Suppose you are driving a 25,000-kg Mercedes down the road at 75 m/s (about 150 mph—you're really booking it!) when a somewhat impatient trucker whose vehicle weighs 80,000 kg rams you from behind at a velocity 250 m/s (he has a pretty impressive truck) in an effort to encourage you to be in a bit more of a hurry. Needless to say, your engines blow up, but the gods of physics are smiling kindly on your fiery plight and are arranging that only a negligible amount of mass is lost in the fireball and flying auto parts. The mangled wreck is a perfectly inelastic collision, so find the final velocity at which you and the truck go careening forward.

Since the collision is perfectly inelastic, the only equation we have is conservation of momentum, with the condition that the final velocities are the same.

$$m_{truck}v_{truck,initial} + m_{car}v_{car,initial} = v_{final}(m_{car} + m_{truck})$$

Here, the truck is one object and the car is another, so rewriting the subscript "truck" as "1" and "car" as "2," we can write

$$m_1 v_{10} + m_2 v_{20} = (m_1 + m_2)v_f \qquad (7\text{-}7)$$

This is the general format by which all inelastic collisions will be solved. Plugging in the numbers that we have, we find that the final velocity is

$$(80000)(250) + (25000)(75) = v_{final}(80000 + 25000)$$
$$21875000 = v_{final}(105000)$$
$$v_{final} = 208 \text{ m/s}$$

But now suppose, in light of your last fiery experience with truckers pushing 500 mph, you attach magnetic deflecting technology to the back of your car so that next time it happens you have a perfectly *elastic* collision. And suppose that, sure enough, having bought a new Mercedes identical to the old one, and cruising along the highway at your same favorite speed of 75 m/s, the same reckless trucker hits you from behind with the same speed as before, and having the same mass as before. Now what are the final velocities of the truck and your car?

For a perfectly elastic collision, both mechanical energy and linear momentum are conserved, so we have two equations to work with—a fortuitous fact, since we have two variables (two final velocities) to solve for.

$$\frac{1}{2}m_{truck}v_{truck,initial}^2 + \frac{1}{2}m_{car}v_{car,initial}^2 = \frac{1}{2}m_{truck}v_{truck,final}^2 + \frac{1}{2}m_{car}v_{car,final}^2 \tag{7-8}$$

and

$$m_{truck}v_{truck,initial} + m_{car}v_{car,initial} = m_{truck}v_{truck,final} + m_{car}v_{car,final}$$

Like before, we can generalize these two equations using numerical subscripts instead of "truck" and "car." Calling the truck object 1 and the car object 2, conservation of energy becomes

$$\frac{1}{2}m_1v_{10}^2 + \frac{1}{2}m_2v_{20}^2 = \frac{1}{2}m_1v_{1f}^2 + \frac{1}{2}m_2v_{2f}^2 \tag{7-9}$$

Likewise, conservation of momentum—the same equation as we used for the inelastic collision, but without the assumption that the final velocity would be the same—becomes

$$m_1v_{10} + m_2v_{20} = m_1v_{1f} + m_2v_{2f} \tag{7-10}$$

It's easiest to start by just plugging in what we know.

$$\tfrac{1}{2}(80000)(250^2) + \tfrac{1}{2}(25000)(75^2) = \tfrac{1}{2}(80000)v_{truck,final}^2 + \tfrac{1}{2}(25000)v_{car,final}^2$$

and

$$(80000)(250) + (25000)(75) = 80000v_{truck,final} + 25000v_{car,final}$$

We can simplify by canceling out the one-halves in the conservation of energy and then doing the arithmetic.

$$5140625000 = 80000v_{truck,final}^2 + 25000v_{car,final}^2$$

$$21875000 = 80000v_{truck,final} + 25000v_{car,final}$$

Note that we are waiting to round out the insignificant figures until the *end* of the problem.

One can proceed by picking a variable to solve for and use substitution. The final velocity of the car works as well as any other.

$$25000v_{car,final} = 21875000 - 80000v_{truck,final}$$

$$v_{car,final} = 875 - 3.2v_{truck,final}$$

Now substituting this into the law of conservation of energy,

$$5140625000 = 80000v_{truck,final}^2 + 25000v_{car,final}^2$$

$$5140625000 = 80000v_{truck,final}^2 + 25000(875 - 3.2v_{truck,final})^2$$

Simplifying by carrying out the square,

$$5140625000 = 80000v_{truck,final}^2 + 25000(765625 - 5600v_{truck,final} + 10.24v_{truck,final}^2)$$

Multiplying the 25,000 through,

$$5140625000 = 80000v_{truck,final}^2 + 1.914 * 10^{10} - 1.4 * 10^8 v_{truck,final} + 256000v_{truck,final}^2$$

Collecting like terms,

$$336000v_{truck,final}^2 - 1.4 * 10^8 v_{truck,final} + 1.4 * 10^{10} = 0$$

We can simplify further and make the numbers more manageable by dividing by 336,000,

$$v_{truck,final}^2 - 416.7\, v_{truck,final} + 4166.7 = 0$$

Now using the quadratic formula,

$$v_{truck,final} = \frac{-b \pm \sqrt{b^2 - 4ac}}{2a}$$

$a = 1$
$b = 416.7$
$c = 4166.7$

$$v_{truck,final} = \frac{416.7 \pm \sqrt{(416.7)^2 - 4(1)(4166.7)}}{2}$$

$$v_{truck,final} = \frac{416.7 \pm \sqrt{156972}}{2}$$

$$v_{truck,final} = \frac{416.7 \pm 396.2}{2}$$

$$v_{truck,final} = 406.5 \text{ m/s or } 10.25 \text{ m/s}.$$

So $v_{car,final} = 875 - 3.2 v_{truck,final} = -425.8$ m/s or 842.2 m/s.

Both sets of answers are mathematically consistent. But there's a physical consideration we need to think about: solid bodies don't pass through each other, a fact that is nowhere reflected in the mathematical equations. If the car were to end up with a velocity of −425.8 m/s and the truck with a velocity of 406.5 m/s, then the truck would have had to have passed through the car, since the truck started *behind* the car and the condition that the collision is perfectly elastic requires that they not smash through each other. Therefore, this solution, while mathematically consistent, is physically irrelevant. The correct answer is that the impatient truck driver will have a final velocity of only 10.25 m/s (a good speed to drive down a neighborhood alley) and the car given a nice boost up to 842.2 m/s (or at least it would, if the sound barrier didn't stop it first!).

Collisions can be one-dimensional or two-dimensional. Both examples worked out here were one-dimensional; for two-dimensional problems, we simply break up the law of conservation of linear momentum into its *x*- and *y*-components. For a perfectly inelastic collision, this is easy—we're usually solving for v_f, and each equation for a two-dimensional perfectly inelastic collision will only have one final velocity variable to be solved for:

$$m_1 v_{10x} + m_2 v_{20x} = (m_1 + m_2) v_{fx} \tag{7-11}$$

$$m_1 v_{10y} + m_2 v_{20y} = (m_1 + m_2) v_{fy} \tag{7-12}$$

Solve for v_{fx} and v_{fy}, and those are the *x* and *y* components of the final velocity. The magnitude and direction can be found using the Pythagorean theorem and inverse tangent as before.

A two-dimensional perfectly *elastic* collision is considerably harder. These will have three equations—conservation of energy, and both the *x*- and *y*-components of conservation of momentum—but *four* variables. Each object involved in the collision is going to have its own final velocity, and we need *two* variables—either the *x*- and *y*-component of each, or the magnitude and direction of each—to give all the information we need about each final velocity. One simply cannot solve three equations for four variables; one of the four variables has to be given.

In most of the homework problems, one of the final angles will be given. This leaves the magnitudes of both final velocities, and one of the angles, left to be solved for. The problem, however, is that the other final angle will appear in one equation as the argument of the *cosine* function, whereas it will appear in the other equation as the argument of the *sine* function. Let's work this through by taking an example.

Suppose a 3 kg ball moving at 12 m/s at a 23° angle hits a 4 kg ball moving at 10 m/s along a −30° angle. If the lighter ball is bounced back to a −26° angle, what will the final velocities of both balls be?

We start by writing down the three equations we need. We'll be using conservation of energy and two equations (x- and y-components) for conservation of momentum:

$$\frac{1}{2}m_1v_{10}^2 + \frac{1}{2}m_2v_{20}^2 = \frac{1}{2}m_1v_{1f}^2 + \frac{1}{2}m_2v_{2f}^2 \qquad (7\text{-}13)$$

$$m_1v_{10x} + m_2v_{20x} = m_1v_{1fx} + m_2v_{2fx} \qquad (7\text{-}14)$$

$$m_1v_{10y} + m_2v_{20y} = m_1v_{1fy} + m_2v_{2fy} \qquad (7\text{-}15)$$

Instead of having magnitudes and components all in the same set of equations, giving us more variables than we need, we'd like to write the x- and y-components in terms of the magnitude and direction of the vectors:

$$v_{1fx} = v_{1f}\cos\theta$$
$$v_{1fy} = v_{1f}\sin\theta$$
$$v_{2fx} = v_{1f}\cos\varphi$$
$$v_{2fy} = v_{1f}\sin\varphi$$

Let's plug in the values we've been given, using 334° as equivalent to −26° and 330° as equivalent to −30°:

$$\frac{1}{2}(3)(12^2) + \frac{1}{2}(4)(10^2) - \frac{1}{2}(3)v_{1f}^2 + \frac{1}{2}(4)v_{2f}^2$$

$$3(12\cos 23) + 4(10\cos 330) = 3(v_{1f}\cos 334°) + 4(v_{2f}\cos\varphi)$$

$$3(12\sin 23) + 4(10\sin 330) = 3(v_{1f}\sin 334°) + 4(v_{2f}\sin\varphi)$$

Conservation of energy simplifies to

$$832 = 3v_{1f}^2 + 4v_{2f}^2$$

while the two conservation of momentum equations simplify to

$$67.8 = 2.7v_{1f} + 4v_{2f}\cos\varphi$$

$$-5.94 = -1.32v_{1f} + 4v_{2f}\sin\varphi$$

Our normal method of solving coupled equations is to pick a variable, solve one equation for it, and substitute into the next equation. These equations are easiest to solve for the angle first. These aren't linear equations, however, and we'll end up solving for the *cosine* or *sine* of the angle rather than going directly to the angle first, and cosine and sine are different functions—you can't simply substitute one for the other. Fortunately, however, the Pythagorean identity relates the two. Taking the Pythagorean theorem,

$$a^2 + b^2 = c^2$$

where a and b are the legs of a right triangle with hypotenuse c,

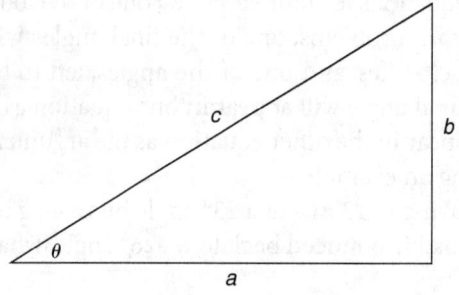

CHAPTER 7 Conservation of Linear Momentum and Collisions **161**

we can divide both sides by c^2, giving us

$$\frac{a^2}{c^2} + \frac{b^2}{c^2} = 1$$

$\frac{a^2}{c^2}$ is the square of the ratio between the adjacent side to the angle and the hypotenuse of the triangle—cosine squared, in other words. Likewise, $\frac{b^2}{c^2}$ is the square of the ratio between the opposite side to the angel and the hypotenuse of the triangle, or sine squared. The Pythagorean identity is therefore

$$\cos^2\theta + \sin^2\theta - 1 \qquad (7\text{-}16)$$

Using this identity, we can rewrite the sine of θ in terms of the cosine:

$$\sin\theta = \sqrt{1 - \cos^2\theta} \qquad (7\text{-}17)$$

The angle we'll be using here is φ. θ is the symbol used in general for the first unknown angle that we need a name for. Let's rewrite conservation of momentum in the y-direction using the Pythagorean identity:

$$-5.94 = -1.32v_{1f} + 4v_{2f}\sqrt{1 - \cos^2\theta}$$

These three equations are easiest to solve for if we start by solving conservation of momentum in the x-direction for φ, and substitute that into conservation of momentum in the y-direction:

$$67.8 = 2.7v_{1f} + 4v_{2f}\cos\varphi$$

$$67.8 - 2.7v_{1f} = 4v_{2f}\cos\varphi$$

$$\frac{67.8}{4v_{2f}} - \frac{2.7v_{1f}}{4v_{2f}} = \cos\varphi \qquad (7\text{-}18)$$

Now we can substitute this expression for $\cos\varphi$ into conservation of momentum in the y-direction. First, however, in order to square $\cos\varphi$, we have to use FOIL:

$$\left(\frac{67.8}{4v_{2f}} - \frac{2.7v_{1f}}{4v_{2f}}\right)^2 = \cos\varphi$$

$$\cos^2\varphi = \frac{287.3}{v_{2f}^2} - \frac{22.8v_{1f}}{v_{2f}^2} + \frac{0.456v_{1f}^2}{v_{2f}^2}$$

One need not worry that the units in each term apparently do not match because each of these constants has units attached to them, units that we have omitted for clarity's sake.

Substituting this expression for $\cos^2\varphi$ in conservation of momentum in the y-direction:

$$-5.94 = -1.32v_{1f} + 4v_{2f}\sqrt{1 - \cos^2\varphi}$$

$$-5.94 = -1.32v_{1f} + 4v_{2f}\sqrt{1 - \frac{287.3}{v_{2f}^2} + \frac{22.8v_{1f}}{v_{2f}^2} - \frac{0.456v_{1f}^2}{v_{2f}^2}}$$

Putting everything under the radical over a common denominator, $\frac{v_{2f}^2}{v_{2f}^2}$, is easy to do and will help simplify:

$$-5.94 = -1.32v_{1f} + 4v_{2f}\sqrt{\frac{v_{2f}^2}{v_{2f}^2} - \frac{287.3}{v_{2f}^2} + \frac{22.8\,v_{1f}}{v_{2f}^2} - \frac{0.456v_{1f}^2}{v_{2f}^2}}$$

Everything in the denominator under the radical is now v_{2f}^2, which is still under a square root sign. Fortunately, the square root of a quantity squared is simply that original quantity, since squaring and taking the square root are inverse functions of each other, so

$$5.94 = -1.32v_{1f} + 4v_{2f}\frac{\sqrt{v_{2f}^2 - 287.3 + 22.8v_{1f} - 0.456v_{1f}^2}}{v_{2f}}$$

Needless to say, v_{2f} in the numerator cancels out v_{2f} in the denominator:

$$5.94 = -1.32v_{1f} + 4\sqrt{v_{2f}^2 - 287.3 + 22.8v_{1f} - 0.456v_{1f}^2}$$

Now we need to solve for one of the final velocities. To do this, we're going to have to get out of that ugly square root sign. This can be easily accomplished by isolated everything under the square root, and then squaring both sides:

$$5.94 + 1.32v_{1f} = 4\sqrt{v_{2f}^2 - 287.3 + 22.8v_{1f} - 0.456v_{1f}^2}$$

$$\frac{5.94}{4} + \frac{1.32v_{1f}}{4} = \sqrt{v_{2f}^2 - 287.3 + 22.8v_{1f} - 0.456v_{1f}^2}$$

$$1.485 + 0.33v_{1f} = \sqrt{v_{2f}^2 - 287.3 + 22.8v_{1f} - 0.456v_{1f}^2}$$

The left side has to be squared using FOIL; the right side is squared simply by removing the square root sign:

$$2.205 + 0.9801v_{1f} + 0.1089v_{1f}^2 = v_{2f}^2 - 287.3 + 22.8v_{1f} - 0.456v_{1f}^2$$

Now we can collect like terms and solve for one of the velocities. Since v_{2f}^2 only appears once, it's easiest to solve for that:

$$v_{2f}^2 = 0.5649v_{1f}^2 - 21.82v_{1f} + 289.5$$

We now have the final velocity of the second object in terms of the final velocity of the first object. The only equation we haven't used yet is conservation of energy, and we can substitute this expression into that equation, replacing v_{2f}^2 and leaving us with a quadratic equation that is only a function of v_{1f}:

$$832 = 3v_{1f}^2 + 4v_{2f}^2$$

$$832 = 3v_{1f}^2 + 4(0.5649v_{1f}^2 - 21.82v_{1f} + 289.5)$$

$$832 = 3v_{1f}^2 + 2.26v_{1f}^2 - 87.28v_{1f} + 1158$$

This needs to be put into quadratic form by collecting like terms in order for the quadratic formula to be employed:

$$5.26v_{1f}^2 - 87.28v_{1f} + 326 = 0$$

Further simplifying by dividing each term by 5.26,

$$v_{1f}^2 - 16.59v_{1f} + 61.98 = 0$$

Now employing the quadratic formula,

$$v_{1f} = \frac{16.59 \pm \sqrt{16.59^2 - 4(61.98)}}{2} = 21.945 \text{ or } -5.355 \, m/s$$

In either case, this velocity will be at a 334° angle. Both answers are mathematically consistent, so let's try them both and see what we get for the other velocity and angle:

$$v_{2f}^2 = 0.5649v_{1f}^2 - 21.82v_{1f} + 289.5$$

$$v_{2f}^2 = 0.5649(21.945^2) - 21.82(21.945) + 289.5 = 82.7$$

$$v_{2f} = 9.09 \text{ m/s}$$

On the other hand, if we try -5.355 m/s for the first velocity,

$$v_{2f}^2 = 0.5649v_{1f}^2 - 21.82v_{1f} + 289.5$$

$$v_{2f}^2 = 0.5649(-5.355)^2 - 21.82(-5.355) + 289.5$$

$$v_{2f} = 20.56 \text{ m/s}$$

For each of these sets of velocities ($v_{1f} = 21.945$ with $v_{2f} = 9.09$, and $v_{1f} = -5.355$ with $v_{2f} = 20.56$), let's find the missing angle φ:

$$\cos\varphi = \frac{67.8}{4v_{2f}} - \frac{2.7v_{1f}}{4v_{2f}}$$

$$\cos\varphi = \frac{67.8}{4(9.09)} - \frac{2.7(21.945)}{4(9.09)} = 0.235$$

$$\varphi = \cos^{-1}(0.235) = 76.4°$$

Taking the other pair of possible velocities:

$$\cos\varphi = \frac{67.8}{4v_{2f}} - \frac{2.7v_{1f}}{4v_{2f}}$$

$$\cos\varphi = \frac{67.8}{4(20.56)} - \frac{2.7(-5.355)}{4(20.56)} = 1.00$$

$$\varphi = \cos^{-1}(1.00) = 0°$$

These two-dimensional perfectly elastic scattering or collision problems have tedious and difficult algebra, and there are only four of them in the homework set. Let's end the chapter by summarizing our method for solving each type of problem.

- For a one-dimensional perfectly inelastic collision, use

$$m_1v_{10} + m_2v_{20} = (m_1 + m_2)v_f$$

Solve for v_f.
- For a two-dimensional perfectly inelastic collision, use

$$m_1v_{10x} + m_2v_{20x} = (m_1 + m_2)v_{fx}$$

$$m_1v_{10y} + m_2v_{20y} = (m_1 + m_2)v_{fy}$$

Solve for v_{fx} and v_{fy} giving you the x- and y-components of the final velocity, and use the Pythagorean theorem to find the magnitude of the final velocity and an inverse trigonometric function (such as inverse tangent) to find the direction.
- For a one-dimensional perfectly elastic collision, use

$$\frac{1}{2}m_1v_{10}^2 + \frac{1}{2}m_2v_{20}^2 = \frac{1}{2}m_1v_{1f}^2 + \frac{1}{2}m_2v_{2f}^2$$

$$m_1v_{10} + m_2v_{20} = m_1v_{1f} + m_2v_{2f}$$

Solve for one of the final velocities in conservation of momentum, and substitute it into conservation of energy.
- For a two-dimensional perfectly elastic collision, use

$$\frac{1}{2}m_1v_{10}^2 + \frac{1}{2}m_2v_{20}^2 = \frac{1}{2}m_1v_{1f}^2 + \frac{1}{2}m_2v_{2f}^2$$

$$m_1v_{10x} + m_2v_{20x} = m_1v_{1fx} + m_2v_{2fx}$$

$$m_1v_{10y} + m_2v_{20y} = m_1v_{1fy} + m_2v_{2fy}$$

Solve conservation of momentum in the x-direction for the cosine of the angle, use the Pythagorean identity to rewrite conservation of momentum in the y-direction to replace the sine of the angle with $\sqrt{1 - sin^2\theta}$ if θ is the angle used, solve conservation of momentum in the x-direction for $\cos\theta$, use FOIL to square it, and substitute into conservation of momentum in the y-direction. Solve conservation of momentum in the y-direction for everything under the radical, squaring both sides by removing the radical on that side and using FOIL on the other. Solve for the velocity that is left standing by itself, and substitute into conservation of energy. Put conservation of energy into quadratic form, and solve for the velocity. Go back and substitute this number for the velocity into expressions already solved for the other velocity and for the cosine of the angle to find the other two variables.

HOMEWORK FOR CHAPTER 7

Name _____

A. A 25,000-kg truck moving at 40 m/s hits a 10,000-kg van moving at 25 m/s from behind. Since this is a one-dimensional, perfectly inelastic collision, what is the final speed of the wreck?

B. A 25,000-kg truck moving at 40 m/s on a road angled at 17° hits a 10,000-kg van moving at 25 m/s on a road angled at 60° at an intersection between the two roads. Since this is a two-dimensional, perfectly inelastic collision, what is the final velocity of the wreck?

C. You have accidentally wandered into Brobdingnag (if you do not remember this name you should read the book it came from again), and the giants are playing ball. A 25,000-kg bowling ball moving at 40 m/s hits a 10,000-kg bowling ball moving at 25 m/s from behind. Assuming a one-dimensional, perfectly elastic collision, what are the final velocities of the two balls?

D. While still in Brobdingnag, a 25,000-kg bowling ball moving at 40 m/s on a path angled at 17° hits a 10,000 kg bowling ball moving at 25 m/s on a path angled at 60°, causing the lighter ball to be deflected off to an 65° angle. Assuming this is a two-dimensional, perfectly elastic collision, what are the final velocities of the balls?

CHAPTER 7 Conservation of Linear Momentum and Collisions

E. You are trying to impress a date with your bowling skills, and since it has been scientifically shown that certain people's IQ scores drop several points in the presence of quite attractive specimens of the opposite gender[http://www.scientificamerican.com/article.cfm?id=why-interacting-with-woman-leave-man-cognitively-impaired], you are making quite the spectacle of both of you at the alley by trying to use physics to improve your performance while imposing creative challenges on yourself to look impressive. Instead of standing at the back of the lane and rolling the ball toward the pins like most people, you place a 5-kg bowling ball at rest at the back of the lane and try to hit it toward the pins with a perfectly elastic collision using a 1-kg ball rolled at 5 m/s from a 25° angle from the horizontal (defining the horizontal as the direction on the floor perpendicular to the lanes). Sure enough, your perfect spin sends the larger ball flying—or at least *moving*—straight down the center of the lane toward the pins. What will the velocity of the larger bowling ball be?

F. In one dimension, a 5-g ball moving to the right at 7 m/s collides with an 8-g ball moving to the left at 4 m/s. The collision is perfectly elastic. Find the final velocities of both balls.

G. In one dimension, a 12-kg ball moving to the right at 9 m/s collides with a 4-kg ball moving to the left at 3 m/s. The collision is perfectly elastic. Find the final velocities of both balls.

H. Your teenage kid finally gets his driver's license. He takes off in your brand-new car (which weighs 2000 kg) down the road at 47 m/s when he hits a parked semi (44,000 kg) and, because the gods of physics are contriving the situation to fit into one dimension, they plow forward in a perfectly inelastic fashion in the same direction that the new car was moving to begin with. Find the final velocity of the wreck.

I. Now it's your other kid's turn behind the wheel. This time you give her the pickup truck, since those things are indestructible. It has a mass of 1250 kg, and she is happily driving at 39 m/s due northeast down a cornfield out in the middle of nowhere in Nebraska when she unfortunately flies over a rock and hits an SUV with mass 1750 kg heading 70° north of east at 100 m/s. The collision is completely inelastic. Find the final velocity of the wreck.

J. You are playing paintball and you see your enemy's reflection in the mirror. The mirror is oriented so that the *y*-axis is perpendicular to it. You are hiding behind a pillar and do not want to come out to shoot at him and expose yourself, so you fire at his reflection in the mirror. The mirror has a mass 1 kg, and the paintball has a mass 0.05 kg. The paintball hits the mirror at a speed of 8 m/s at an angle of 25° from the normal (from the axis coming out of the mirror). The mirror is pushed backward upon the impact, remaining facing the same direction it was before. Assuming that the collision was perfectly elastic (with real paintballs it won't be—this is something you can verify at home), find the final velocity of the paintball. Does it hit your enemy?

K. You are a physicist, and you are wondering what would happen if you fired a 9.109 * 10^{-31} kg particle (say, a positron) moving at 20 m/s in the x direction off of a 1.672 * 10^{-24} kg particle (a proton fits this description pretty well) moving at 12 m/s at 110° from the x-direction. A blip on your detector shows that the lighter particle hit the screen at an angle of 100° from the x direction (quite conveniently, you are simplifying matters by performing the experiment in Flatland, where you do not have to worry about that pesky third dimension). Find the final velocities of both particles. Assume a classical (non-relativistic), perfectly elastic collision.

Rotational Kinematics and Dynamics

Rotational Motion

As impressive—and elegantly beautiful—as the framework of physics that we've developed in the first seven chapters is, it suffers from a major limitation. It, strictly speaking, only applies to motion in a straight line. The principle of inertia says that, given the absence of a force acting on the object, an object in motion will stay in motion, and an object at rest will stay at rest—and any change in *direction* is considered to be an acceleration. "Natural" motion is, so to speak, in a straight line—and in Euclidean geometry, a straight line is the shortest distance between two points, and anyone having doubts about this can prove it using the calculus of variations. (It turns that the universe as a whole, just like the surface of the earth, is not flat—its geometry *isn't* Euclidean. But on small scales, it's flat enough that we won't notice any difference. Ask your professor about manifold theory, which explains how curved objects can appear flat.)

One major wrinkle making circular motion difficult is that the acceleration is nonuniform. We've assumed in almost all of our problems that the acceleration was going to be constant—we haven't even told you a name for the change of acceleration versus time. (The name, for those who are curious, is "jerk.") It turns out to be quite difficult to describe the acceleration of circular motion as a function of time in Cartesian coordinates. Instead, we turn to *polar* coordinates, which instead of using the coordinates x, y, and z use the coordinates r, θ (an angle), and ϕ (another angle). Polar coordinates are a student's headache in their own right, and we're not going to use them too rigorously. We're especially not going to take the polar components of vectors given the magnitudes. We've already seen them though—Newton's universal law of gravitation had a unit vector "$\hat{r}$," which is a vector pointing in the direction between the objects exerting forces on each other. That law was already written in polar coordinates, even though one may not have realized it at the time. Problem G10 in Chapter 6 made the claim that in a conservation of energy problem, the *radial component* of the acceleration (the component going in the **r** direction, a polar component) was always $a_r = \dfrac{v^2}{R}$, an acceleration called "centripetal acceleration." This acceleration has a constant magnitude if the *speed* of the object is constant, but the direction of the acceleration (which is still, like always, a vector) will be changing at a constant rate, meaning that we have a nonuniform acceleration. (The expression holds true regardless of the gravitational force acting on the object, because the gravitational force will cause the speed to change—and then we have the nightmare that the magnitude of the radial component of the acceleration is changing, too.)

Even if the magnitude of the acceleration is constant, its direction will not be. Life is easier when acceleration is constant, and this is going to motivate our definition of a whole new set of *angular* quantities from which we'll rederive the four kinematics equations from scratch.

Linear kinematics was developed earlier in the book from two simple concepts: position (x) and time (t). Time will remain fundamental and unchanged in angular motion, but instead of dealing with *position*, it will be convenient to use the *angle traversed* as our fundamental quantity instead of displacement. In order to be careful and rigorous, let's give a formal definition of what we're calling the "angle" traversed. The angle traversed in a circular path is defined as

$$\theta \equiv \frac{s}{r} \qquad (8\text{-}1)$$

where s is the *path length* (**not** displacement) of the portion of the circle traversed, and r is the radius of the circle.

An astute observer should raise two objections immediately. The first is that the quantity given is a scalar—we don't know whether we've gone clockwise around the circle or counterclockwise. The second is that the arc length s has units of meters, and so does the radius r—meaning that the angle θ *has to be unitless*, clearly a false statement when the angle is measured in degrees!

For this reason, this definition only works when the angle θ is measured in the pseudounit "radians." Radians are not a unit—in dimensional analysis they can be dropped or added as we please, without needing to cancel them out with anything else. A radian is *defined* by the definition given above—one radian is the angle traversed when the path length is equal to the radius. This means we can be given a conversion factor between radians and degrees: π radian = 180°. But it's still not a true unit—the "angle" θ is the quotient between two lengths, and is therefore unitless. It's really a ratio, not an angle, but it's a ratio that corresponds directly and linearly to the angle that produces that ratio, and for this reason we report the ratio as having the pseudounit "radians."

The first objection was that this ratio "angle" θ is a scalar, and we need to have a vector in order to know which direction our displacement is. The answer is given by a simple convention called the "right-hand rule." Curl your fingers of your right hand around the direction of motion while keeping your thumb pointed out; your thumb points in the "direction" of the vector $\boldsymbol{\theta}$. Furthermore, if this motion or displacement is counterclockwise (in which case your thumb should be pointing toward you), the direction is considered to be "positive," since the final angle minus the initial angle will be positive (angles are measured rotating counterclockwise from the x axis of a Cartesian coordinate system). If the motion is clockwise and your thumb is pointed away from you, the direction is "negative."

Let's go ahead and define quantities analogous to velocity and acceleration, using $\boldsymbol{\theta}$ instead of $\mathbf{x}$, and then use the same reasoning that we did before to derive the four kinematics equations.

Displacement was defined as final position minus initial position, so we can define angular displacement as

$$\Delta\boldsymbol{\theta} \equiv \boldsymbol{\theta}_f - \boldsymbol{\theta}_i \qquad (8\text{-}2)$$

Velocity was defined as displacement over time, so *angular velocity* is similarly defined as

$$\boldsymbol{\omega} \equiv \frac{\Delta\boldsymbol{\theta}}{t} \qquad (8\text{-}3)$$

Angular acceleration is defined as

$$\boldsymbol{\alpha} \equiv \frac{\Delta\boldsymbol{\omega}}{t} \qquad (8\text{-}4)$$

Angular acceleration is a quantity that isn't going to change. Its direction is given by the right-hand rule, so the fact that an object is moving in a circle *doesn't* imply a varying angular acceleration. This is our motivation for dealing with angular acceleration instead of the old quantity **a**.

These are definitions for *average* angular velocity and acceleration, of course—in the case that one needs *instantaneous* angular velocity and acceleration, one would as always take the limit as $t \to 0$, turning the deltas into differentials and making angular acceleration the time derivative of angular velocity, and angular velocity the time derivative of angular displacement.

The Rotational Analogues for the Kinematics Equations

Let's apply the same logic to these quantities that we did with $\Delta \mathbf{x}$, $\mathbf{v}$, and $\mathbf{a}$ to derive the original four kinematics equations, in order to derive four *new* kinematics equations for rotational motion.

$$\alpha \equiv \frac{\Delta \omega}{t}$$

Since $\Delta \omega = \omega_f - \omega_0$, this definition can be rewritten

$$\alpha \equiv \frac{\omega_f - \omega_0}{t}$$

Following the same convention as before and writing the measured value as simply ω rather than ω_f,

$$\alpha \equiv \frac{\omega - \omega_0}{t}$$

Now multiply both sides by t.

$$\alpha t = \omega - \omega_0$$

And then to find ω, add ω_0 to both sides of the equation:

$$\omega = \omega_0 + \alpha t \tag{8-5}$$

which is analogous to our first kinematics equation. Let's derive a rotational analogue for our second kinematics equation.

$$\omega \equiv \frac{\Delta \theta}{t}$$
$$\Delta \theta = \omega t$$

As before, assuming uniform angular acceleration, the average angular velocity is going to be half of the initial angular velocity plus the final angular velocity:

$$\omega_{ave} = \frac{1}{2}(\omega + \omega_0)$$

and so

$$\Delta \theta = \frac{1}{2}(\omega + \omega_0)t \tag{8-6}$$

which is the rotational analogue for the second kinematics equation.

To derive the rotational analogue for the third kinematics equation, we substitute ω into the second rotational kinematics equation:

$$\Delta \theta = \frac{1}{2}(\omega + \omega_0)t$$

$$\Delta \theta = \frac{1}{2}(\omega_0 + \alpha t + \omega_0)t$$

$$\Delta\theta = \frac{1}{2}(2\omega_0 + \alpha t)t$$

$$\Delta\theta = \frac{1}{2}(2\omega_0 t + \alpha t^2)$$

$$\Delta\theta = \omega t + \frac{1}{2}\alpha t^2$$

Now remembering the definition

$$\Delta\theta \equiv \theta - \theta_i$$

(dropping the subscript for the final angle, as conventional), we see that

$$\theta - \theta_i = \omega_0 t + \frac{1}{2}\alpha t^2$$

Or

$$\theta = \theta_i + \omega_0 t + \frac{1}{2}\alpha t^2 \qquad (8\text{-}7)$$

This is the rotational analogue to the third kinematics equation.

To find the fourth kinematics equation, we substituted time into the third kinematics equation. The first rotational kinematics equation solved for time gives us

$$t = \frac{\omega - \omega_0}{\alpha} \qquad (8\text{-}8)$$

and substituting this for time in equation 8-7 gives us

$$\theta = \theta_0 + \omega_0 \frac{\omega - \omega_0}{\alpha} + \frac{1}{2}\alpha\left(\frac{\omega - \omega_0}{\alpha}\right)^2$$

The simplification takes the some process as it did before. A student can easily verify that this gives us

$$\omega^2 = \omega_0^2 + 2\alpha\Delta\theta \qquad (8\text{-}9)$$

What if one wants to convert back and forth between the angular and linear quantities defined? Well, tangential speed is the path length divided by time—remember that *speed* is path length over time rather than displacement over time. So,

$$v = \frac{\Delta s}{t} \qquad (8\text{-}10)$$

But we defined the ratio θ such that

$$\theta \equiv \frac{s}{r} \qquad (8\text{-}11)$$

Or

$$s = \theta r \qquad (8\text{-}12)$$

Assuming that the radius r of the circle is constant, this gives us velocity as

$$v = r\frac{\Delta\theta}{t}$$

But $\frac{\Delta\theta}{t}$ is ω. So

$$v = r\omega \qquad (8\text{-}13)$$

The *tangential* component of the acceleration (the change in *speed* irrespective of direction) is $\frac{\Delta v}{t}$ where again we are assuming that r is not changing, so

$$a_t = r\frac{\Delta \omega}{t}$$

where $\frac{\Delta \omega}{t}$ is of course α. So

$$a_t = r\alpha \tag{8-14}$$

The centripetal component of the acceleration was given earlier as

$$a_c = \frac{v^2}{r} \tag{8-15}$$

which is an equation not easily derivable without calculus. However, we can substitute $r\omega$ for v in order to expression the centripetal acceleration as

$$a_c = r\omega^2 \tag{8-16}$$

Energy in Rotational Systems

In Chapter 6, we found that complicated kinematics problems can be simplified greatly by introducing the concept of work and using the work-kinetic energy theorem and conservation of total mechanical energy as a substitute for the kinematics equations (or, in simple cases, as a shorthand for the fourth kinematics equation). We'd like to preserve our freedom to use energy to solve problems with rotational motion. Since energy is a scalar—which means the fact that a rotating object is changing direction is therefore irrelevant—the conversion is simple. We just take the definition of kinetic energy and substitute an expression involving angular velocity for linear velocity.

We found above that $v = r\omega$, so the kinetic energy of a rotating body is simply this expression substituted into $KE = \frac{1}{2}mv^2$. Making this substitution gives us $KE = \frac{1}{2}m(r\omega)^2 = \frac{1}{2}mr^2\omega^2$.

Angular velocity is in many ways the rotational analogue to linear velocity, and we want the kinetic energy of a rotational body to be equal to half times some rotational analogue to mass times the angular velocity squared, in order to make the format perfectly analogous to linear kinetic energy. As we can see, mr^2 is that rotational analogue to mass that makes the formulae have the same appearance. We call mr^2 the **moment of inertia** of a rotating body, and denote it by the symbol "I," giving us the equation

$$KE_R = \frac{1}{2}I\omega^2 \tag{8-17}$$

Usually in problems that involve rotational motion the object in question will actually be a collection of particles or a solid body, rather than a point particle going in a circle (which is a rather narrow and limited application of this physics). Applying the principle of superposition, the kinetic energy of a *collection* of particles is the sum of the kinetic energies of the individual particles:

$$KE_R = \frac{1}{2}\left(\sum_i m_i r_i^2\right)\omega^2 \tag{8-18}$$

where the moment of inertia is consequently defined as

$$I \equiv \sum_i m_i r_i^2 \tag{8-19}$$

In this case r_i is the distance from each particle to the axis of rotation. In the case of a continuous extended object, we view it as an infinite number of discrete particles, and the summation sign becomes an integral over the volume of the object.

Take careful note that the rotational kinetic energy KE$_R$ *only* applies to motion in a circle. Most motion in reality has both translation and rotational motion—for example, a rolling barrel—and we need to take this into consideration when using the law of conservation of energy. In such situations, the total kinetic energy must be written as

$$KE = \frac{1}{2}mv^2 + \frac{1}{2}I\omega^2 \qquad (8\text{-}20)$$

where v is now the purely *translational* component of the velocity. (For example, we may be given a wheel or a barrel rolling at a certain velocity v. The center of the barrel or wheel is moving forward at v, and from v we can calculate ω, and then use *both* terms added together as the kinetic energy.)

Calculating the moments of inertia of extended objects requires calculus, and the moment of inertia depends on where the axis of rotation is. Instead of going through the calculus, we'll present the results in a table and accept the results (which are useful for problems, even problems that themselves do not require the use of calculus) on faith. The moments of inertia presented are technically the moment of inertia for the *center of mass*, which is the average position of the object. Since physics usually involves determining the position of an object, we model all objects as point particles with precise positions determined by the center of mass.

Object	Moment of Inertia
Solid sphere rotating on its axis:	$I = \frac{2}{5}MR^2$

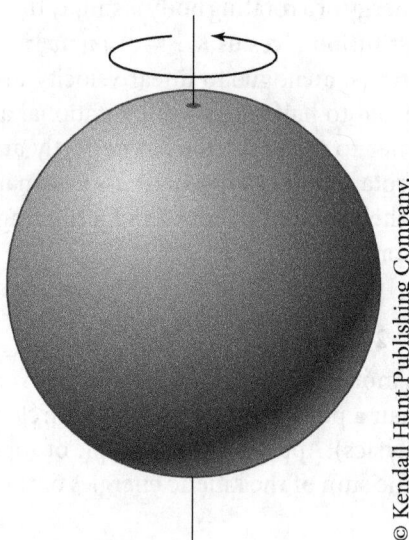

© Kendall Hunt Publishing Company.

Thin spherical shell rotating on its axis:

$$I = \frac{2}{3}MR^2$$

Solid cylinder rotating on its axis:

$$I = \frac{1}{2}MR^2$$

Hoop or thin cylindrical shell rotating on its axis:

$$I = MR^2$$

Hollow cylinder with inner radius R_1 and outer R_2:

$$I = \frac{1}{2}M(R_1^2 + R_2^2)$$

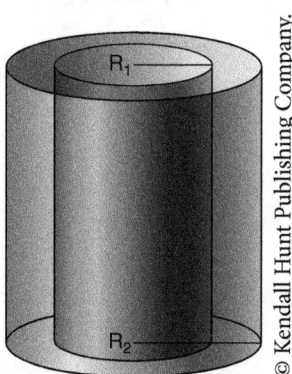

Rectangular plate with sides a and b:

$$I = \frac{1}{12}M(a^2 + b^2)$$

Long thin rod rotating around center:

$$I = \frac{1}{12}ML^2$$

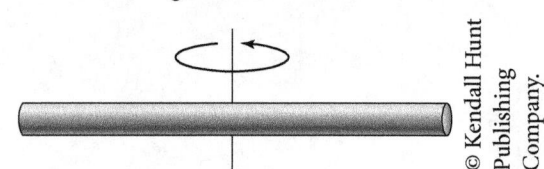

Long thin rod rotating around end:

$$I = \frac{1}{3}ML^2$$

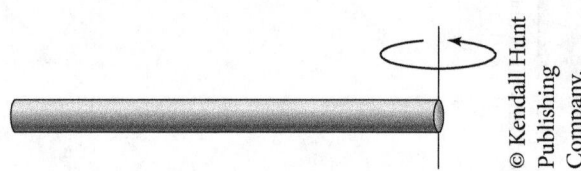

For the sphere, spherical and cylindrical shells, and the cylinder, M refers to the total mass and R to the radius. For the long thin rod, M refers to the mass and L to the length. The hollow cylinder with two radii and the rectangular plate are presumed to be rotating around the axis in the center.

All but the last equation had the objects rotating around an axis passing through the **center of mass** of the object, the average position or "middle." It is often useful to find the moment of inertia of an object rotating around something other than the center of mass. To do this, we apply the **parallel axis theorem**, which requires calculus for its derivation. The conclusion of this theorem, which for our purposes will simply be taken on faith, is that the moment of inertia of an object rotating around an axis a distance D from and parallel to an axis going through the center of mass is given by

$$I = I_{CM} + MD^2 \tag{8-21}$$

To take an example, let's take the moment of inertia for a long thin rod rotating around an axis going perpendicular to the rod and through the center of mass, and use the parallel-axis theorem to find the moment of inertia for the same rod rotating around the end of the rod. The table above gives the original moment of inertia as

$$I = \frac{1}{12}ML^2$$

so the parallel-axis theorem gives the moment of inertia for the same rod rotating around an axis a distance $\frac{L}{2}$ away from the center as being

$$I = \frac{1}{12}ML^2 + M\left(\frac{L}{2}\right)^2 = \frac{1}{12}ML^2 + M\frac{L^2}{4} = \frac{1}{12}ML^2 + M\frac{3L^2}{12} = \frac{4}{12}ML^2 = \frac{1}{3}ML^2$$

exactly as claimed in the table above.

Force in Rotational Systems

Next, we wish to find a rotational analogue for force. This is being presented after energy, contrary to the order in which energy and force were presented for translation motion. This is because we are already familiar with the concept of energy, and solving problems using Newton's second law with rotational systems is considerably more difficult.

The magnitude of a force is given by $F = ma$, and, not surprisingly in a rotational system, it is convenient to rewrite this by substituting an expression containing α for the acceleration. We are given above that the tangential or translational component of the acceleration is

$$a_t = r\alpha$$

which gives Newton's second law as being

$$F_t = mr\alpha$$

Previously we defined the moment of inertia I as the rotational analogue for mass, so we would like to write this equation with I instead of m. Using $I = mr^2$ for the simplest case, a point particle, we can solve for mr and find that

$$\frac{I}{r} = mr$$

which is a useful substitution to make in Newton's second law. Doing so shows that

$$F_t = \frac{I}{r}\alpha$$

and multiplying both sides by r in order to get $I\alpha$ by itself on the right-hand side (so that it looks analogous to ma in the translational version of Newton's second law), we find that

$$F_t r = I\alpha$$

The quantity $F_t r$ is the rotational analogue for force, and is called **torque** and denoted by the Greek letter tau:

$$\tau = I\alpha \tag{8-22}$$

It is time to address a protest that the astute purist should have raised by now. We've been simplifying the calculations in a gross way by treating r as a scalar, just the magnitude of the distance from the force to the axis of rotation without worrying about the direction, but of course displacement **r** *is* a vector, and our assumption of circular motion has required that **r** be always pointing from the object to its center of motion. You *can't* use a vector **r** in our calculations though, because we divided by r in order to show how Newton's second law for torque can be derived from the translation version of Newton's second law, **F** = m**a**. Division is an operation that isn't defined for vectors, *period*. One might have suspected that I pulled the wool over your eyes, and that is precisely what I did.

The equation $F_t r = I\alpha$ *only* holds true for circular motion in which the direction of motion is always perpendicular to the radius. The quantity $F_t r$ is a vector, and the tangential force is always perpendicular to the radius. The wool was pulled over our eyes when the claim was made that this quantity is called torque, τ. It is the torque—but we can define the torque in a more general manner, for motions that are not necessarily circular and for forces where we do not know (or are too lazy to calculate) the tangential component.

F and **r** are both vectors, and it is these two vectors that must be multiplied together. However, the only way we've defined vector multiplication—the dot product—gives us a *scalar* as a result, and torque is quite clearly a vector (it must be, since the quantity on the other side of the equation—$I\alpha$—is a vector, and common sense dictates that if two quantities are really equal, are really the *same*, whatever is true of one must be true of the other. A vector cannot be equal to a scalar; they are fundamentally different sorts of things. (Actually, for subtle and ignorable reasons we'll find out in Chapter 9, torque is actually a *pseudovector* or "axial vector" because it is antisymmetric rather than symmetric under inversion, but let's not make things more complicated than we need to, as fascinating as the mathematics actually is.)

It turns out there is a second definition for multiplication between vectors which is an operator completely different from the dot product, called the *cross product*. More detail about the cross product will be given in Chapter 9 when we have to start becoming more careful with our mathematics; for the purposes of the present chapter, two facts suffice in order to give the student facility in using the cross product:

1. The magnitude of the cross product **A** × **B** between two vectors **A** and **B** separated by an angle θ is given by $AB \sin \theta$, where A and B are the magnitudes of the vectors **A** and **B** respectively. Note that for perpendicular vectors, such as F_t and **r**, $\theta = 90°$ and $\sin 90° = 1$, which is why we were justified using r instead of **r** in the expression $F_t r$.
2. The direction of **A** × **B** is given by the right-hand rule, a rule that has at least three different variants. The easiest one to remember for finding the direction of a cross product is to point your thumb up and your index finger straight out, the way a child might when pretending to point a gun, and then your middle finger (the one used for certain impolite gestures) to the left, perpendicular to the other two fingers. If the thumb points in the direction of **A**, and the index finger in the direction of **B**, then the middle finger points in the direction of the cross product. The disadvantage to using this rule is that it can only be used when **A** and **B** are perpendicular to each other, or when the components of **A** and **B** are taken one at a time (to give the respective components of cross product).

The torque is actually defined as

$$\boldsymbol{\tau} \equiv \boldsymbol{r} \times \boldsymbol{F}. \tag{8-23}$$

$\tau = I\alpha$ is proven from this, using reasoning going backward from the way it was presented here. Simplicity of understanding motivated the treatment of the topic going in the direction we did—taking the assumption of circular motion and tangential force—rather than a more general treatment using the more difficult concept of torque as a cross product. The simplifying assumption we used was pedagogically useful, but $\tau = I\alpha$ can be used for *any* force and *any* acceleration. (If the motion is purely translational, of course, the *angular* acceleration will be 0 because r will be 0, and therefore so will the torque, which is the only reason we've been able to solve problems without using $\tau = I\alpha$ up until now.)

For an object in equilibrium, the acceleration **a** and therefore the angular acceleration $\boldsymbol{\alpha}$ will both be 0. While the linear two-dimensional problems in Chapter 5 required two equations to solve—Newton's second law for the x- and y-components of the force separately—problems involving rotational motion are generally or at least sometimes going to require *three* equations to solve, the two we had used before and $\tau = I\alpha$:

$$\sum \tau = I\alpha$$

$$\sum F_x = ma_x \tag{8-24}$$

$$\sum F_y = ma_y \tag{8-25}$$

One further wrinkle is going to make these problems even harder: we generally aren't going to be given the axis of rotation for any given torque. Objects in nature simply do not come with "axis of rotation" labels glued to the bottom of them, yet we still will want to know how any given force is going to cause the object to rotate. Fortunately, these equations *do not specify any particular axis of rotation*. We can choose any axis of rotation we wish, and doing so determines both I and α. It is usually easiest to choose an axis of rotation going through the center of mass. For an extended object of uniform mass distribution, the center of mass will be the position right in the center (for nonuniform mass distribution, calculus is required). For a collection of point particles, center of mass is the "average position," given by

$$x_{CM} = \frac{\sum_i m_i x_i}{\sum_i m_i} \tag{8-26}$$

Let's do an example. Suppose that a heavy plank with mass 350 kg and dimensions 1 m × 10 m is leaning against a cardboard wall (which weighs 5 kg, is 50 m long, 0.5 m thick, and 20 m tall) at an angle 50°. If the coefficient of static friction between the cardboard wall and the ground is 0.8 and a man 2 m tall and 5 m from the wall is pulling on the plank with a rope to a point 2 m from the top of the plank, what is the minimum tension in the rope needed to keep the cardboard wall from falling over?

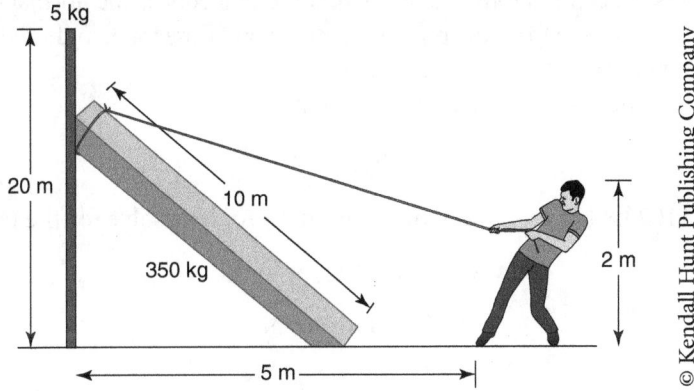

What we are really asking here is what tension needs to be applied to the plank so that the torque from the normal force between the plank and the wall is equal to the normal force of the friction between the ground and the wall keeping it from tipping over. Let's choose an axis of rotation going through the center of mass of the brick wall, for simplicity's sake. In this case the frictional torque between the wall and the ground is

$$\tau = \mu_s N h \tag{8-27}$$

where $\mu_s = 0.8$, the normal force N equals the weight of the wall (5 kg *9.8 m/s^2), and the height of the center of mass from the ground (the ground being where the frictional force is applied) is half the height of the wall, or 10 m. We can just use the magnitudes here because the frictional force is applied in the horizontal direction, perpendicular to the wall. So the torque applied by friction will be 0.8*5*9.8*10 = 392 N m.

If a light object is leaning against a flimsy wall, the friction from the ground will keep the wall from tipping. So if we want the *minimum* tension needed to keep the wall from tipping, friction will be helping us, causing a torque on the wall in the positive or counterclockwise direction, while the normal force between the plank and the wall cause a torque on the wall in the negative or clockwise direction; meanwhile, the tension in the rope and the weight of the plank will cause torques *on the plank* (for which we can take the axis of rotation to be the point where it touches the ground, so that the normal force between the ground and the plank will not cause any torque on the plank since $r = 0$) rather than on the wall. The torques caused by the weight of the wall and the normal force between the ground and the wall will cancel each other out.

Because the plank is tilted 50° up from the ground and the weight acts straight down, the angle θ is going to be the angle complementary to 50°, which is 90° − 50° = 40°. The distance r from the force to the axis of rotation is *half* the length of the plank, since the center of mass is in the middle of the plank—so $r = 5$ m. Therefore, the torque applied by the weight of the plank is 350 kg * 9.8 m/s² * 5 m * sin 40° = 11023.8 N m. The torque caused by the tension is applied 2 m from the axis of rotation, and if the man is standing at a height 2 m and the rope is attached to a point on the plank 8 sin 50° m high, the rope will be attached to the plank on a position 6.13 m above where he is holding it and 5 m across from where he is holding it, giving it an angle $\tan^{-1}(5/6.13) = 39.2°$ from the plank. So the torque caused by the tension T is $T * 8 \sin 39.2° = 5.06 * T$ N m.

We have to be careful here: The tension on the rope causes a torque *on the plank*, and the weight of the torque causes a torque on the plank, while the normal force of the plank on the wall and the frictional force of the ground on the wall cause torques *on the wall*. The connecting factor is that the principle of superposition dictates that torque caused by the normal force of the plank on the wall will be equal and opposite to the *total* torque on the plank, since the torque on the plank will cause it to rotate with an angular acceleration proportional to the torque it transfers to the wall. The angular acceleration the plank receives from its weight and the tension in the rope causes it to fall into the wall in such a way as to act like a torque of its own, and that torque is the normal force of the plank *on* the wall times the distance from the point of application of the force to the axis of rotation. The total torque on the plank is $5.06T - 11023.8$ N m.

So Newton's second law for torques tells us that

$$\sum \tau = 0$$

or $5.06T - 11023.8$ N m = 392 N m. All we have to do is solve for the tension, T.

$$5.06T = 11023.8 + 392$$

$$5.06T = 11415.8 \text{ N m}$$

$$T = 2.26 * 10^3 \text{ N}$$

Rotational Power and Momentum

Finally, rotational analogues for power and momentum can be developed. Power is defined as

$$P = \frac{W}{t} \tag{8-28}$$

and the work-kinetic energy theorem tells us that W = ΔKE, regardless as to whether this kinetic energy is rotational or translational. So for a rotating body, assuming a constant moment of inertia,

$$P = \frac{\Delta KE}{t} = \frac{\frac{1}{2}I\omega^2}{t} = \frac{I\omega^2}{2t} \tag{8-29}$$

We can solve $\tau = I\alpha$ for I to get $I = \frac{\tau}{\alpha}$ and substitute it into the equation above to obtain

$$P = \frac{\tau \omega^2}{2t\alpha} \tag{8-30}$$

Using the fact that $\alpha \equiv \dfrac{\Delta\omega}{t}$, the power it takes to move an object from rest to a final angular velocity ω is

$$P = \frac{\tau\omega^2}{2t} * \frac{t}{\omega} = \frac{\tau\omega}{2} \qquad (8\text{-}31)$$

which gives the *average* power delivered by a torque moving an object from rest to an angular velocity ω. Assuming that we're starting from rest with uniform acceleration, the *average* power is average of the final *instantaneous* power and the initial *instantaneous* power (where $\omega = 0$). Since the initial *instantaneous* power is 0 (because $\omega = 0$), the final *instantaneous* power is twice the average power, or $P = \tau\omega$. A derivation of this equation using calculus will verify this result.

Finally, *angular momentum* is found by starting with the definition of linear momentum as

$$\boldsymbol{p} \equiv m\boldsymbol{v} \qquad (8\text{-}32)$$

Substituting $r\omega$ for v gives

$$\boldsymbol{p} = mr\omega$$

Using $I = mr^2$ to rewrite m in terms of I,

$$m = \frac{I}{r^2}$$

$$\boldsymbol{p} = \frac{I}{r^2}r\omega$$

$$\boldsymbol{p} = \frac{I}{r}\omega \qquad (8\text{-}33)$$

I is the rotational analogue of m and $\boldsymbol{\omega}$ is the rotational analogue of $\boldsymbol{v}$, so the rotational analogue of $\boldsymbol{p}$ should be $I\boldsymbol{\omega}$. This can be isolated by multiplying both sides by r:

$$\boldsymbol{p}r = I\boldsymbol{\omega}$$

Of course, the purist will object—*r* is a vector, and when we're not doing shady manipulations like dividing by its magnitude while ignoring its vector properties entirely, we need to remember that it is a vector. Everything we did was actually completely rigorous—we used the definition of moment of inertia as $I = mr^2$ (ignoring the fact that it only applies to a point particle without any *internal* spin, which fine—linear momentum was only defined for point particles ignoring internal structure too), and r^2 is in fact a scalar, **r***r. But what this equation ($\boldsymbol{p}r = I\boldsymbol{\omega}$) *implies* is that angular momentum ($\boldsymbol{p}r$) is going the same direction as $\boldsymbol{p}$—all we did was scale it by the length of the radius. The momentum, of course, is $m\boldsymbol{v}$ where $\boldsymbol{v}$ is the translational velocity, going *perpendicular* to **r**. The operation that multiplies vectors together preserving the fullest magnitude when the two vectors are perpendicular is the cross product, so not surprisingly angular momentum is defined as

$$\boldsymbol{L} \equiv \boldsymbol{r} \times \boldsymbol{p} \qquad (8\text{-}34)$$

giving us

$$\boldsymbol{L} = I\boldsymbol{\omega} \qquad (8\text{-}35)$$

Using Newton's second law for rotational motion, $\boldsymbol{\tau} = I\boldsymbol{\alpha}$, to solve for I gives us

$$I = \frac{\tau}{\alpha}$$

Substituting this into the expression for angular momentum yields

$$\boldsymbol{L} = \frac{\tau}{\alpha}\boldsymbol{\omega}$$

Recalling that $\alpha \equiv \dfrac{\omega}{t}$ and assuming an initial angular velocity of 0 (which would be the initial angular velocity for which the torque τ would cause an angular acceleration α up to an angular velocity ω corresponding to a change in angular momentum ΔL),

$$\Delta L = \dfrac{\tau}{\frac{\omega}{t}}\omega$$

$$\Delta L = \tau * t$$

$$\tau = \dfrac{\Delta L}{t} \tag{8-36}$$

just as

$$F = \dfrac{\Delta p}{t} \tag{8-37}$$

Consequently, when no torque is acting on an object (or when there is no angular acceleration), $\Delta L = 0$. This gives us our third conservation law: conservation of angular momentum. Let's close the chapter by reviewing the three conservation laws:

- *Mechanical energy* is conserved when no nonconservative forces are acting on an object.
- *Linear momentum* is conserved when no net external force is acting on an object.
- *Angular momentum* is conserved when no net external torque is acting on an object.

HOMEWORK FOR CHAPTER 8

Name _____

A. Is the angular acceleration vector for a clock positive or negative?

B. Which direction are the velocity vectors at points A and B? Let y be the vertical direction, x be horizontal, and z coming out of the paper.

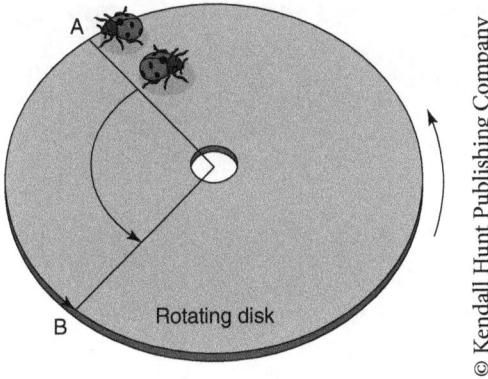

C. Consider a long thin rod with length L, like this:

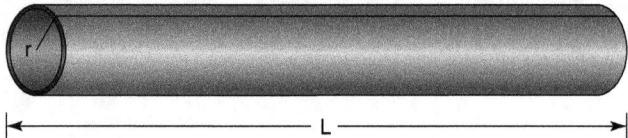

The moment of inertia is ML²/3. Draw the axis of rotation.

D. You have a crane boom with a mass of 10,000 kg being held up by a cable and holding a weight of 2500 kg.

1. Find the torque caused by the weight.

2. Find the torque caused by the cable.

3. Find the torque caused by any other force holding it in equilibrium.

E. You and your friends are engaging in your favorite pastime: rolling irregularly shaped blobs down the hill. The hill has an incline of 0.50 radians (drawing may not be exact, the artist was missing his protractor!), and the blob is rolling around its center of mass, helpfully pointed out to you by the black dot. When the blob is at a height of 0.5 m, its center of mass has a velocity of 4 m/s. What is the acceleration of the center of mass at that point? Find the magnitudes of the acceleration and velocity vectors.

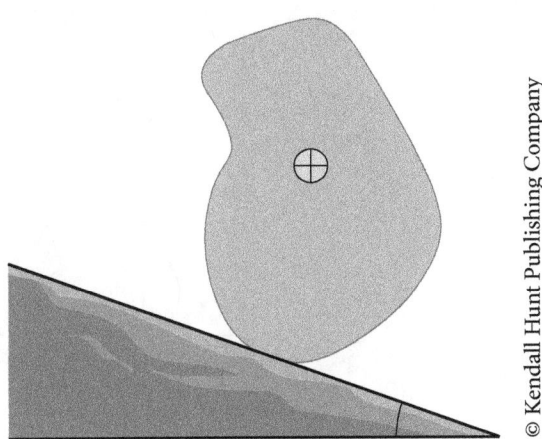

F. A rolling barrel with initial velocity 20 m/s encounters a hill 10 m tall to slow it down:

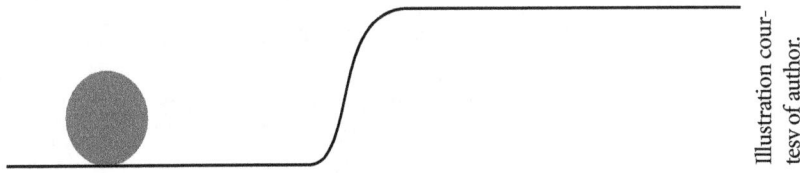

1. State where (if anywhere) its angular speed will be 0.

2. State where (if anywhere) its rotational kinetic energy will be 0.

3. State where (if anywhere) its translation kinetic energy will be 0.

4. For each of the two flat surfaces (on top of the hill and on the bottom), compare the translational kinetic energy, rotational kinetic energy, and potential energy (i.e., state which quantities are greater than or less than which other quantities).

G. Your favorite steroid-fueled baseball player is at the plate about to hit another home run. Which direction is the angular momentum vector of the baseball bat? Let y be up, x be to the right, and z be out of the board.

H. A figure skater, spinning in a circle and dazzling the Olympic judges, brings her arms together for a final flourish.
   1. Does her angular momentum increase, decrease, or stay the same after she pulls her arm in?

   2. Does her angular velocity increase, decrease, or stay the same after she pulls her arm in?

   3. Does her angular acceleration increase, decrease, or stay the same after she pulls her arm in?

I. Since three opinions are better than one, you take up the habit of always carrying three wristwatches. One of them is synchronized with your lab's atomic clock. Another is running low on battery and is 3 minutes behind, while the third you wound up a little too much and is running 6 minutes ahead. All of them were set to the same time when you purchased them. List the comparative angular velocities of the three watches in order of greatest to smallest.

J. A turntable with radius 12 cm is being explored by two bugs, one at 5 cm from the center and another, very dizzy beetle at 2 cm from the center. The turntable is spinning at 7 rad/s. Which bug will have the greater angular momentum?

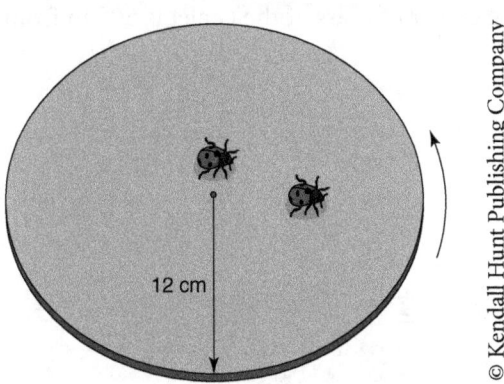

K. A ladder of mass 20 kg and length 5 m can be held up against a wall because the wall exerts a normal force against it. The grass has a coefficient of static friction of 0.35. Find the normal force of the wall against the ladder.

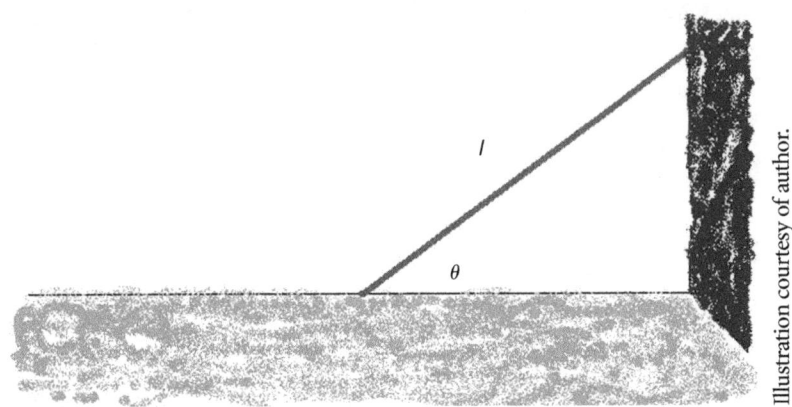

L. You are a sucker for scenic overlooks, so you want to build a house overhanging a cliff—literally. The way you do this is by taking a 20,000 N concrete beam and hang it over the edge of the cliff, with one edge at the cliff ledge and the other edge 700 m away in free space. You plan to hold it up with a sturdy cable, which can withstand a maximum tension 500,000 N, and which you tie to a tree that is 200 m tall, weighs 2,000 N and is held to the ground with a force of 45 N. The tree is 7 m from the edge of the cliff. Assume that the tree is genetically engineered to be really tough, so that if the cable pulls too hard the whole tree will come out instead of breaking in the middle. Your house is a beautiful Swiss chalet that weighs 750,000 N, and the drop to the ground at the bottom of the cliff is a mere 2,000 m. The center of mass of the chalet is 500 m from the ledge of the cliff.

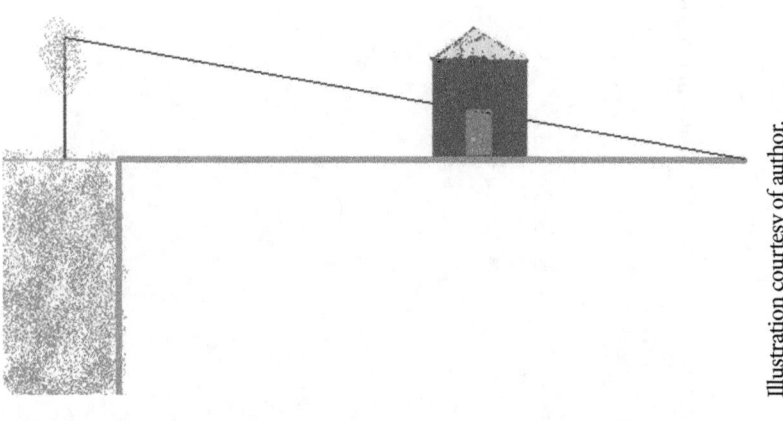

Illustration courtesy of author.

1. Find the normal force of the cliff on the beam. Give a quantitative value.

2. What is the tension in the cable?

3. Will this scheme work?

M. Instead of spending money on a new table when you move into your first apartment, you try instead to balance a wooden board on two upright wooden pegs. The board has a length 5 m and a weight 25 N. It overhangs the left peg by 1 m and the right peg by 2 m. Determine the normal force that each peg exerts on the board.

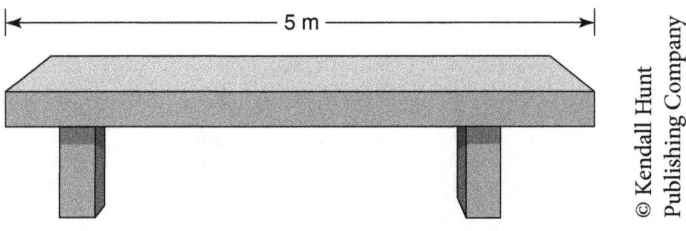

N. A pirate captain has asked you to help him engineer a new plank to have prisoners walk overboard from. He needs the plank to fall overboard with the prisoners in order not to leave any evidence behind when the Navy comes to storm his ship. He has a uniform beam of length 8 m and mass 100 kg is resting on two pivots. One pivot is at the left end, and the other is 5 m from the left end. The prisoner will walk from the left end toward the right. Find where the prisoner will be when the plank tips. Assume the prisoner weighs 72 kg.

O. One can determine a person's center of mass by having her recline on a plank resting on two scales, as shown in the figure. A person 2.00 m tall lies on such a plank, and the scale near her head reads 405 N and the one near her feet reads 360 N. Find her center of mass.

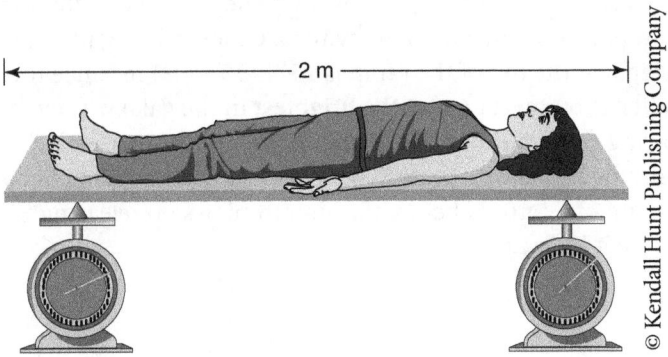

P. A planet is in orbit around the center of mass of a binary star system. The planet has a mass of $4.32*10^{23}$ kg, the larger star has a mass of $7.89*10^{37}$ kg, and the smaller star a mass of $2.01*10^{33}$ kg. The figure is not quite to scale (since the bigger star is about ten thousand times the size of the smaller one, and the smaller star about ten billion times the size of the planet), but at the point in the planet's orbit, find the quantitative value of the gravitational force on the planet, and find the center of mass of the system.

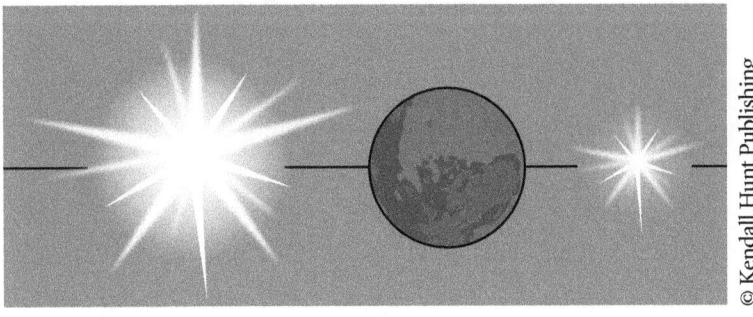

Q. Aliens have made their first contact with earth and are planning to make a visit. In order to greet them, your government sets up a 35-m tall ramp angled 50° from the horizontal in order for their spacecraft to land on, ending on a spring-operated weight scale that reports weight proportional to the compression of the spring (so your government can take some surreptitious data about the spacecraft and its technology). The scale has a spring constant 50,000 N/m (that's particularly high—we want the spring to be sensitive so the aliens don't notice the landing pad drop significantly when they land and thereby realize they're being measured) and a height 25 m above the ground. (So the height at the top of the ramp is 25 + 35 = 60 m, since the base of the ramp is 25 m high.) Unfortunately, the aliens you meet are not the brightest in the galaxy. They forget to give their spacecraft any mechanism for slowing down besides friction from their encounter with the atmosphere. Their spacecraft happens to be a simple barrel with mass 200 kg and radius 5 m. When it comes crashing down to the scale, the scale reads a weight of $7 * 10^6$ N before the aircraft blows up releasing 25,000 J of heat. How fast was the spacecraft going before it blew up?

Coulomb's Law

Vector Arithmetic

We've introduced the concept of vectors in previous chapters, using them to solve two-dimensional kinematics and mechanics problems. However, we've largely avoided the issue of actually adding vectors together or performing elaborate computations with vectors, since we've separated vector equations by components. With electrostatics, we no longer have the luxury of taking this easy route, so we must begin with a discussion of vector computation.

For computation, we are still going to have to break up a vector into its *components*—the magnitude of the vector going along the x direction and the magnitude of the vector going along the y direction.

As we've seen before, a vector can be denoted by the Cartesian coordinate where it would end if the base were placed at the origin (e.g., a vector could be (5,5) if it starts at the origin and points to the location (5,5)), or by its total magnitude and direction (e.g., $5\sqrt{2}$ @ 45°), or as the sum of two orthogonal component vectors (e.g., $5\mathbf{x} + 5\mathbf{y}$).

For addition and subtraction of vectors, one must add or subtract the components of the vector separately. The sums or differences obtained are the components of the resultant vector. For example, the vector $(7, 3) + (8, 4) = (7 + 8, 3 + 4) = (15, 12)$.

Multiplication and division is more tricky. Division is not defined for vectors, so we will not be doing it at all. A vector multiplied by a scalar has its direction unchanged, and its magnitude is multiplied by that scalar quantity (the magnitude is "scaled", in other words). There are two different ways by which we can multiply two vectors by each other, the *dot product* or inner product and the *cross product* or outer product.

The Dot Product

The dot product is found by multiplying the x-components of the vectors together and adding it to the product of the y-components and to the product of the z-components (if the vector has three dimensions). For example, $(5, 8) * (4, 9) = 5 * 4 + 8 * 9 = 20 + 72 = 92$. Note that the result of the dot product is a **scalar**, although it is the product of two vectors.

Another way to calculate the dot product is to multiply the two magnitudes together and multiply by the *cosine* of the angle between them. For example, $(5 @ 45°) * (7 @ 0°) = 35 \cos 45° = 35/\sqrt{2}$.

The Cross Product

The cross product is more complicated because it is a vector, not a scalar. To find the magnitude of the cross product, multiply the magnitudes of the two vectors together and multiply by the *sine* of the angle between them. For example, the magnitude of $(5@45°) \times (7@0°) = 35 \sin 45° = 35\sqrt{2}$. However, the cross product itself is a vector, and has a direction that needs to be computed. To find the components of cross product $\mathbf{a} \times \mathbf{b}$, one must take the determinant of the following matrix:

$$\begin{vmatrix} \hat{x} & \hat{y} & \hat{z} \\ a_x & a_y & a_z \\ b_x & b_y & b_z \end{vmatrix}$$

The determinant is the following combination of sums and products: $\mathbf{x}(a_y b_z - a_z b_y) - \mathbf{y}(a_x b_z - a_z b_x) + \mathbf{z}(a_x b_y - a_y b_z)$. If the two vectors $\mathbf{a}$ and $\mathbf{b}$ lie in the x-y plane, then the cross product will only have a $\mathbf{z}$-component, orthogonal to both $\mathbf{a}$ and $\mathbf{b}$.

A quick way to find the direction of the cross product is the right-hand rule, mentioned in the previous chapter. Point your thumb straight up, your index finger straight ahead, and your middle finger to the left. If your thumb is pointed in the direction of $\mathbf{a}$ and your index finger in the direction of $\mathbf{b}$, then your middle finger points in the direction of the cross product between the two. This only works on your right hand (left hand gets the cross product pointing backwards). The right-hand rule only works when the two vectors are perpendicular to each other; when they are not, we must take the determinant of the matrix.

The cross product is *anticommutative*. This means that unlike regular multiplication, $\mathbf{A} \times \mathbf{B} = -\mathbf{B} \times \mathbf{A}$. Switching the order of the terms multiplied together changes the sign of the answer: it's antisymmetric under inversion. (Technically, this makes the cross product a "pseudovector" or *axial vector*, since true vectors or *polar vectors* are supposed to be symmetric under inversion. This means that many of the "vectors" we saw earlier like torque and angular momentum are actually pseudovectors, and the equations we derived setting them equal to true vectors—like $L = I\omega$ and $\tau = I\alpha$—are technically incorrect, *the way they have been defined* with I as a scalar, since they claim that an axial vector is equal to a polar vector, a contradiction in terms. The reason for the apparent contradiction is that I is not a scalar at all, but a tensor, or matrix that rotates the same way a vector does—and the tensor product between that tensor and those vectors gives a pseudotensor of the first rank, otherwise called a pseudovector. The fact that I is a tensor is reflected in our use of the terminology "moment of inertia"—a "moment" is the diagonal component of a tensor.)

Electrostatics: Coulomb's Law

The **electric force** between two objects is an effect of electric charge. Electric charge comes in two forms—"positive" and "negative" charges, names assigned arbitrarily by Benjamin Franklin. Opposite charges exert an attractive force toward each other, while like charges—whether they are both positive or both negative—exert a repulsive force against each other. Just as another attractive force between two bodies, gravity, is proportional to the product of the two masses and inversely proportional to the distance between them squared (a property of three-dimensional space),

$$F_G = G \frac{M_1 M_2}{r^2} \hat{r} \qquad (9\text{-}1)$$

electric force is also proportional to the two charges (analogous to gravitational mass—one could more fundamentally think of mass as a gravitational "charge") and inversely proportional to the difference between them squared. The proportionality constant, of course, is different.

$$F_e = k_e \frac{q_1 q_2}{r^2} \hat{r} \tag{9-2}$$

For reasons that will become evident when we derive Gauss's law, and also for reasons that will become apparent when we present Maxwell's equations, we will usually want to rewrite the constant k_e as

$$k_e = \frac{1}{4\pi\varepsilon_0} \tag{9-3}$$

where ε_0 is called the "permittivity of free space" and has the value $8.85 * 10^{-12}$ C²/N m². C is the fundamental unit of electrical charge, a "Coulomb" (after the physicist Charles Coulomb), N is the unit of force Newtons, and m is meters. Using this form, Coulomb's law can be rewritten

$$F_e = \frac{1}{4\pi\varepsilon_0} \frac{q_1 q_2}{r^2} \hat{r} \tag{9-4}$$

The electrical constant k_e was measured experimentally by Charles Coulomb, and was found to be equal to $8.99 * 10^9$ N m²/C². Historically, ε_0 was defined by its relationship to k_e. (As a side note for the curious, today it is defined by Maxwell's equations, in terms of another constant called the "permeability of free space," μ_0, which is defined as being exactly $4\pi * 10^{-7}$ T m/A where T is "Tesla," the fundamental unit of magnetism, and A is "Ampere," the fundamental unit of current, such that

$$\varepsilon_0 \equiv \frac{1}{\mu_0 c^2} \tag{9-5}$$

where c is the speed of light in a vacuum, a quantity that must be experimentally determined—however, this is getting far afield of our material, and we will get to it in due time, after we have learned about magnetism.)

q_1 and q_2 in Coulomb's law denote a measurement of the electrical charge of an object. Electrical charge is a fundamental property of nature whose basic unit is called the "Coulomb." The most commonly known charged particles are the proton, which has a positive charge (of $1.6022 * 10^{-19}$ C—the smallest isolated charge found in nature), and the electron, which has a negative charge of $-1.6022 * 10^{-19}$ C. A macroscopic, physical object can only be "charged" (given an electric charge after it was neutral) through the transfer of electrons from one atom (or molecule, or metallic lattice) to another. Protons are very tightly bound within the nucleus, and the only way for an atom to gain or lose a proton is for a nuclear reaction to take place (hence the term *nuclear reaction*—this is the same sort of reaction that was used in the two atomic bombs that ended World War II).* So a positively charged substance is one in which some of the electrons have been removed, leaving an overall positive charge from the protons in the nucleus (but with the same number of protons as a neutral atom would have), and a negatively charged substance is one in which extra electrons have been added to the atoms.

Unlike most of the macroscopic classical forces discussed in Physics I—such as friction, tension, and normal force—which require contact between the two objects for a force to act, the electrostatic force acts between particles at a distance r from each other. (*Electrostatic* means that the particles are not moving. Once a particle starts moving we have to start worrying about magnetism, and the physics becomes a little bit more complicated.) In other words, the electrostatic force is not a "contact" force. How then does one particle exert a force on another without touching it? The simplest answer is that every charged particle has what we call an electric *field* surrounding it. An electric field is a property of the space surrounding a

---

*All nuclear reactions have the same sort of intensity, although nuclear power plants that provide your electricity harness it on a much smaller and more controlled scale, just as your body harnesses combustion [fire] in a much smaller and more controlled scale to give you energy.

charged particle (really, the space of the whole universe, but its strength is proportionate to the force the particle would exert—so it tapers off to nothing at very large distances). The electric field of a particle with charge q_1 is defined as

$$E = \frac{F}{q_2} \qquad (9\text{-}6)$$

where q_2 is the charge of the particle *not* causing the field. In other words, a charged particle (say, with charge q_1) creates an electric field in the space around it, and when another charged particle with charge q_2 enters it, particle 2's charge is multiplied by the electric field in order to get the force acting on it. The electric field is the force felt *by* a particle divided by that particle's charge, since it is a property of space that could affect any charge at that location.

One quick note about the terminology q_1 and q_2: The numbers 1 and 2 are simply used to indicate that these are two different particles we are talking about. One could just as easily name them q_5 and q_6. In fact, many problems will have multiple charges involved, which may be called q_1, q_2 and q_3, for example. If the problem asks for the charge on q_3, one of the two charges in Coulomb's law will be called q_3—one cannot simply multiply $q_1 * q_2$ as they have been defined for that problem. The subscripts and labels are arbitrary. The electrical force on a particle is always proportionate to the charge of *that* particle times the charge of the particle causing the force.

What do you do if you have multiple charges and want to know the force they cause? Electrical force is just like any other force, and is subject to Newton's laws of motion and all the other properties of mechanics that we learned earlier. In particular, Newton's first law states that the net force on an object is the sum of all the individual forces acting on it (also called the principle of superposition). So if we have three charges named q_1, q_2, and q_3, and we want to know the force on q_3, it will be the sum of the force from q_1 and the force from q_2. In other words,

$$F_{on\,3} = k_e \frac{q_1 q_3}{r_{1\,and\,3}^2} + k_e \frac{q_2 q_3}{r_{2\,and\,3}^2}$$

Here $r_{1\,and\,3}$ is the distance between q_1 and q_3, and $r_{2\,and\,3}$ is the distance between q_2 and q_3.

Let's do an example problem to see how we use Coulomb's law in mechanics problems. A typical problem might be that we are given two charges: a $+5$ Coulomb charge at x = 0 and a $+2$ C charge at x = 1, and we want to place a negatively charged particle with charge -3C somewhere in-between the two so that the three particles are in equilibrium. Where does the particle get placed?

We approach this problem the same way we would any other mechanics problem. We have multiple charges and therefore multiple forces acting on the negatively charged particle whose location we want to determine, and Newton's second law tells us what the forces on a particle do:

$$\Sigma \mathbf{F} = m\mathbf{a} \qquad (9\text{-}7)$$

We want the particle to be in equilibrium, which means that its motion is not changing. (We take a frame of reference in which they are at rest. Remember, the Galilean principle of relativity states that there is no preferred "frame of reference" in which something is absolutely at rest—if two objects are moving at constant velocity relative to each other, then each one can say that they are at rest and it is the other one that is moving. Motion—real motion—is therefore defined not by velocity but by acceleration.) To say that its motion is not changing means that its acceleration $a = 0$. So for any object in equilibrium

$$\Sigma \mathbf{F} = 0. \qquad (9\text{-}8)$$

So far so good—the left-hand side of the equation simply means "the sum of all the forces," which we write by adding together all of the electric forces acting on our negatively charged particle. There are two forces acting on it, both of which are given to us by Coulomb's law:

$$k_e \frac{(5\,C)(-3\,C)}{r_1^2} - k_e \frac{(2\,C)(-3\,C)}{r_2^2} = 0$$

Why did we write them as being subtracted from each other? After all, Newton's second law tells us to add.

Well, remember that a force is a vector, although we are only dealing with one dimension here, and that in the x direction it is defined as positive if it is moving in the $+x$ direction and negative if it is moving in the $-x$ direction. So the sign of the force must be positive if it pulls the particle to the left and negative if it pulls the particle to the right.

There are two different conventions for deciding which sign to use, and both of them work. Here's the one I just used, which employs two rules to follow:

1. Give q as the value of the charge of the particle, including its sign (positive value for positive charges, negative value for negative charges).
2. If it acts on a force to the right of it, you add the force. If it acts on a force to the left of it, you subtract the force.

A different convention is used by many textbooks. It employs four rules or more depending on how you want to count them, making it harder to follow and use:

1. Plug in the *absolute value* of the charge for q in Coulomb's law—in other words, make them all positive.
2. Then look and see whether the force between them is between like charges or opposite charges. If they are like charges, it is repulsive; if they are opposite charges, it is attractive.
3. If the charge is repulsive, then
    a. If it is acting on a particle to the right of it, you add the force (make it positive).
    b. If it is acting on a particle to the left of it, you subtract the force (make it negative).
4. If the charge is attractive, then
    a. If it is acting on a particle to the right of it, you subtract the force (make it negative).
    b. If it is acting on a particle to the left of it, you add the force (make it positive).

By either method, in our example problem, the two forces need to be subtracted from each other.

$$k_e \frac{(5\,\text{C})(-3\,\text{C})}{r_1^2} - k_e \frac{(2\,\text{C})(-3\,\text{C})}{r_2^2} = 0$$

This equation has two unknown variables: r_1 and r_2. You cannot solve a single equation with two unknown variables; we remember from algebra that if there are two unknown variables, we must have two equations. The first equation in a problem like this is usually a fundamental law of physics (like conservation of kinetic energy or linear momentum or in this case, Newton's second law); the second equation is usually called a *boundary condition* or *constraint* and is unique to the problem. In this problem, the constraint is that the 5 C charge is placed at $x = 0$ and that the 2 C charge is placed at $x = 1$. The whole system is in equilibrium, meaning that nothing is accelerating, so these charges are fixed there. Since the -3C particle is placed between them, the distance from the negatively charged particle to the particle on its left plus the distance from the negatively charged particle to the particle on its right equals the total distance between the two positively charged particles; in other words, $r_1 + r_2 = 1$ m. (To phrase it more simply, placing the negatively charged particle between the other two simply divides the 1 meter span into r_1 and r_2.) We can solve the constraint equation for one of the variables—say, r_2—and plug it into Newton's second law to get a single equation with a single variable.

$$r_1 + r_2 = 1\,\text{m}$$

$$r_1 = 1\,\text{m} - r_2$$

So the first equation we have for this problem will now look like

$$k_e \frac{(5\,\text{C})(-3\,\text{C})}{(1\,\text{m} - r_2)^2} - k_e \frac{(2\,\text{C})(-3\,\text{C})}{r_2^2} = 0$$

Let's solve it by adding the second term to both sides of the equation.

$$k_e \frac{(5\,C)(-3\,C)}{(1\,m - r^2)^2} = k_e \frac{(2\,C)(-3\,c)}{r^{2^2}}$$

We notice that k_e appears in every term on both sides of the equation, so if we divide both sides by k_e it cancels out everywhere:

$$\frac{k_e}{k_e} \frac{(5\,C)(-3\,C)}{(1\,m - r^2)^2} = \frac{k_e}{k_e} \frac{(2\,C)(-3\,C)}{r^{2^2}}$$

$$\frac{(5\,C)(-3\,C)}{(1\,m - r_1)^2} = \frac{(2\,C)(-3\,C)}{r_1^2}$$

Now we want to get rid of the things in the denominator by multiplying both sides of the equation by those terms. Start by multiplying both sides of the equation by r_1^2:

$$r_1^2 \frac{(5\,C)(-3\,C)}{(1\,mr_1)^2} = \frac{(2\,C)(-3\,C)}{r_1^2} r_1^2$$

$$(2\,C)(-3\,C) = \frac{(5\,C)(-3\,C)}{(1\,m - r_1)^2} r_1^2$$

Now multiply both sides of the equation by $(1\,m - r_1)^2$:

$$(1\,m - r_1)^2 (2\,C)(-3\,C) = \frac{(5\,C)(-3\,C)}{(1\,m - r_1)^2} r_1^2 (1\,m - r_1)^2$$

$$(1\,m - r_1)^2 (2\,C)(-3\,C) = (5\,C)(-3\,C) r_1^2$$

Now multiply everything together:

$$(-6\,C^2)(1\,m - r_1)(1\,m - r_1) = (-15\,C^2) r_1^2$$
$$(-6\,C^2)(1\,m^2 - 1\,m{*}r_1 - 1\,m{*}r_1 + r_1^2) = -15 r_1^2 C^2$$
$$(-6\,C^2)(1\,m^2 - 2\,m{*}r_1 + r_1^2) = -15 r_1^2 C^2$$
$$-6\,C^2 m^2 + 12 r_1 C^2 m - 6\,C^2 r_1^2 = -15 r_1^2 C^2$$

Add the term on the right side to both sides of the equation to collect like terms:

$$-6\,C^2 m^2 + 12 r_1 C^2 m + 9\,C^2 r_1^2 = 0$$

Divide by the unit C^2 since it appears in every term:

$$-6\,m^2 + 12 r_1 m + 9 r_1^2 = 0$$

Note that this is a quadratic equation of r_1. We can use the quadratic formula,

$$r_1 = \frac{-b \pm \sqrt{b^2 - 4ac}}{2a}$$

where $a = 9$, $b = 12\,m$, and $c = -6\,m^2$. (Note that I am assiduously preserving and treating each unit as an algebraic factor instead of taking a shortcut and leaving them out.)

$$r_1 = \frac{-12\,m \pm \sqrt{12^2 m^2 - 4(9)(-6\,m^2)}}{2(9)}$$

$$r_1 = \frac{-12\,m \pm \sqrt{144\,m^2 + 216\,m^2}}{18}$$

$$r_1 = \frac{-12\text{ m} \pm \sqrt{360\text{m}^2}}{18}$$

$$r_1 = \frac{-12\text{ m} \pm \sqrt{36 * 10\text{m}^2}}{18}$$

$$r_1 = \frac{-12\text{ m} \pm 6\text{ m}\sqrt{10}}{18}$$

$$r_1 = \left(\frac{-12 \pm 6\sqrt{10}}{18}\right)\text{m}$$

Again m is the unit meters, so we get an answer in meters, which is good because we are talking about length. There are two solutions to the quadratic equation; let's look at each one separately:

$$r_1 = \frac{-12 + 6\sqrt{10}}{18}\text{m} = 0.38\text{ m}$$

Just for kicks, let's look at the second solution:

$$r_1 = \left(\frac{-12 - 6\sqrt{10}}{18}\right)\text{m} = -1.72\text{ m}.$$

The question specified that we want to place the particle in-between the charge at $x = 0$ m and the one at $x = 1$ m, so a solution that has us place it to the left of both is irrelevant for this problem.

As we can see, even a deceivingly simple-looking one-dimensional problem with two particles can become rather complicated, giving us a quadratic equation. Let's look at a simple two-dimensional application. Electrostatic force, like any other force, is a vector, and to take the sum of two vectors one must add each component separately to find the components of the resultant vector and then add the components in quadrature (i.e., use the Pythagorean theorem to find the hypotenuse between the x-component and the y-component). So let's say, for example, that you have a charge $+5$ C at $x = 0$ m and -4 C at $x = 0.3$ m, both along the x-axis $(y = 0)$. What is the electric field at the point $(0, 0.4\text{ m})$, or 0.4 m up the y-axis?

Our recipe is simple: Find the x-component of the electric field and find the y-component of the electric field. The x-component will only have a contribution from the 4 C charge, since the 5 C charge is on the y-axis directly beneath the point we are looking at. The magnitude of the electric field from the 4 C charge at the point (0, 0.4) is $E = k_e \frac{4}{0.3^2 + 0.4^2}$, using the Pythagorean theorem to find the distance from (0.3, 0) to (0, 0.4). This is the total contribution to the electric field caused by this charge; the x-component will be this value times the cosine of the angle, the cosine of the angle being $\frac{0.3}{\sqrt{0.3^2 + 0.4^2}}$. So $E_x = k_e \frac{4}{0.3^2 + 0.4^2} * \frac{0.3}{\sqrt{0.3^2 + 0.4^2}} = 8.64 * 10^{10}$ N/C.

The y-component of the total electric field will be the y-component of the contribution from the 4 C charge added to the total field from the 5 C charge. The y-component of the contribution from the 4 C charge is $E_x = k_e \frac{4}{0.4^2 + 0.4^2} * \frac{0.3}{\sqrt{0.3^2 + 0.4^2}} = 1.15 * 10^{11}$ N/C replacing the cosine of the angle with the sine, and the contribution to the electric field from the 5 C charge is $5k_e/(0.4^2) = 2.81 * 10^{11}$ N/C, so the total $E_y = 1.15 * 10^{11} + 2.81 * 10^{11} = 3.96 * 10^{11}$ N/C.

The magnitude of the total electric field then will be $E = \sqrt{E_x^2 + E_y^2} = 9.51 * 10^{11}$ N/C.

Now let's do a problem involving Newton's laws and electric forces in two dimensions. Say you have two particles, each of charge q and mass m, hanging from ropes of length L. They will each hang at an angle $\pm\theta$ from the horizontal:

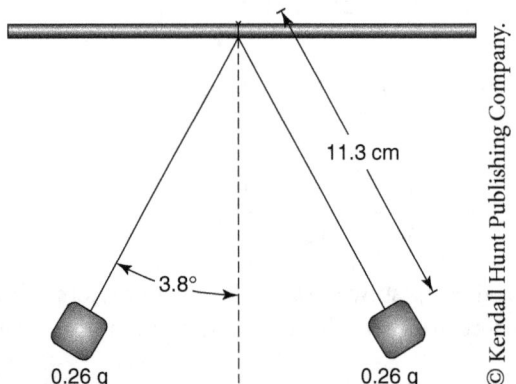

The masses, length of the string, and angle could be different. Let's derive the value of the charge first using the numbers, and then in general.

Take the charge on the left and draw a force diagram. The force diagram will have the y-component of tension pulling it up and the gravitational force pulling it down. The angle here is measured from the vertical, so the y-component of tension will be the side *adjacent* to the angle divided by the hypotenuse, times the total magnitude: In other words, we use cosines now for vertical components and sines for horizontal components. So $\Sigma F_y = 0$ gives us $T\cos\theta - mg = 0$.

In the horizontal direction, electrostatic repulsion will push the charge to the left (exerting a negative force, since the left is the negative x direction), and the x-component of the tension will pull it in the positive direction. We stated that both of the charges are the same, and the distance from each charge to the dotted line in the middle is $0.113\ m * \sin 3.8° = 0.0075\ m = 77\ mm$. So $\Sigma F_x = 0$ gives us $T\sin\theta - kq^2/(2*0.0075)^2 = 0$, or $T\sin 3.8° - 3.99*10^{13} q^2 = 0$.

In order to solve for q, we need to know what T is, so going back to the equation for the x-components,

$$T\cos 3.8° = mg$$

$$0.998T = (0.00026)(9.8)$$

$$T = 0.00255\ N$$

So $0.00255 \sin 3.8° - 3.99 * 10^{13} q^2 = 0$.

$$3.99 * 10^{13} q^2 = 1.69 * 10^{-4}$$

$$q^2 = 4.24 * 10^{-18}$$

$$q = 2.06 * 10^{-9}\ C = 2.06\ nC$$

Now let's derive a general expression for q without giving values for L, θ, and m. We can find an expression either in terms of L or in terms of L $\sin\theta$ = "a," which is more convenient to write. The distance between the two particles is either r = 2 L $\sin\theta$ or r = 2a. We'll use the latter notation here.

As before, $T\sin\theta - kq^2/4a^2 = 0$ and $T\cos\theta - mg = 0$. Let's solve for q.

$$T\sin\theta = kq^2/4a^2$$

$$T\cos\theta = mg$$

Divide the first equation by the second one:

$$\text{Tan } \theta = kq^2/(4a^2 mg)$$

$$q^2 = 4a^2 mg \tan \theta / k$$

$$q = 2a\sqrt{\frac{mg \tan \theta}{k_e}}$$

Or, using $k_e = \dfrac{1}{4\pi\varepsilon_0}$,

$$q^2 = (4a^2 mg \tan \theta)(4\pi\varepsilon_0)$$

$$q^2 = 16a^2 mg \pi\varepsilon_0 \tan \theta$$

$$q = 4a\sqrt{mg\pi\varepsilon_0 \tan \theta}$$

Finding the Electric Field of a Dipole

A dipole is a configuration of two charges of equal magnitude but opposite sign spaced very close together. Advanced physics problems use the electric field and potential from a dipole to approximate the field or potential from an arbitrary charge distribution, so it's useful to get practice dealing with them now. Let's say you have a dipole composed of charge q and charge $-q$ separated by a distance d, aligned along the z-axis with the positive charge on top. We want to find the electric field at some large distance "z" along the z-axis above the midpoint between the two charges.

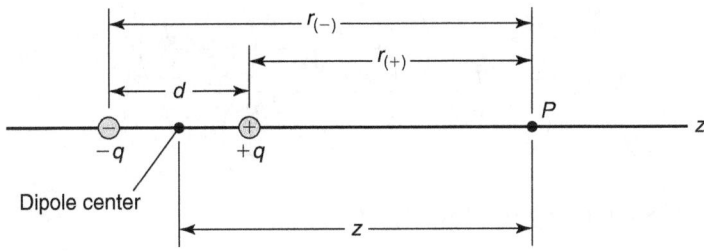

Let's work it out. The field is just the field we obtain from Coulomb's law adding the contributions from the two charges together. We'll call the field from the positive charge E_+ and the field from the negative charge E_-.

$$E = E_+ + E_- = \frac{1}{4\pi\varepsilon_0}\frac{q}{r_+^2} + \frac{1}{4\pi\varepsilon_0}\frac{-q}{r_-^2} \qquad (9\text{-}9)$$

r_- is $z - \frac{1}{2}d$, and r_+ is $z + \frac{1}{2}d$.

$$E = \frac{1}{4\pi\varepsilon_0}\frac{q}{\left(z - \dfrac{d}{2}\right)^2} + \frac{1}{4\pi\varepsilon_0}\frac{-q}{\left(z + \dfrac{d}{2}\right)^2} \qquad (9\text{-}10)$$

Simplify by factoring out $\dfrac{q}{4\pi\varepsilon_0}$, and multiplying both the numerator and denominator by $\dfrac{1}{z^2}$:

$$E = \frac{q}{4\pi\varepsilon_0 z^2}\left(\frac{1}{\left(1 - \dfrac{d}{2z}\right)^2} - \frac{1}{\left(1 + \dfrac{d}{2z}\right)^2}\right) \qquad (9\text{-}11)$$

Now put both fractions over a common denominator:

$$E = \frac{q}{4\pi\varepsilon_0 z^2}\left(\frac{\left(1+\frac{d}{2z}\right)^2}{\left(1-\frac{d}{2z}\right)^2\left(1+\frac{d}{2z}\right)^2} - \frac{\left(1-\frac{d}{2z}\right)^2}{\left(1+\frac{d}{2z}\right)^2\left(1-\frac{d}{2z}\right)^2}\right)$$

Using a property of exponents,

$$E = \frac{q}{4\pi\varepsilon_0 z^2}\left(\frac{\left(1+\frac{d}{2z}\right)^2}{\left(\left(1-\frac{d}{2z}\right)\left(1+\frac{d}{2z}\right)\right)^2} - \frac{\left(1-\frac{d}{2z}\right)^2}{\left(\left(1+\frac{d}{2z}\right)\left(1-\frac{d}{2z}\right)\right)^2}\right)$$

Collecting like terms,

$$E = \frac{q}{4\pi\varepsilon_0 z^2}\frac{\left(1+\frac{d}{2z}\right)^2 - \left(1-\frac{d}{2z}\right)^2}{\left(\left(1-\frac{d}{2z}\right)\left(1+\frac{d}{2z}\right)\right)^2}$$

Simplifying the denominator

$$E = \frac{q}{4\pi\varepsilon_0 z^2}\frac{\left(1+\frac{d}{2z}\right)^2 - \left(1-\frac{d}{2z}\right)^2}{\left(1+\frac{d}{2z}-\frac{d}{2z}-\frac{d^2}{4z^2}\right)^2}$$

$$E = \frac{q}{4\pi\varepsilon_0 z^2}\frac{\left(1+\frac{d}{2z}\right)^2 - \left(1-\frac{d}{2z}\right)^2}{\left(1-\frac{d^2}{4z^2}\right)^2}$$

Expanding and simplifying the numerator,

$$E = \frac{q}{4\pi\varepsilon_0 z^2}\frac{1+\frac{d}{2z}+\frac{d}{2z}+\frac{d^2}{4z^2} - \left(1-\frac{d}{2z}-\frac{d}{2z}+\frac{d^2}{4z^2}\right)}{\left(1-\frac{d^2}{4z^2}\right)^2}$$

$$E = \frac{q}{4\pi\varepsilon_0 z^2}\frac{1+\frac{d}{2z}+\frac{d}{2z}+\frac{d^2}{4z^2} - 1+\frac{d}{2z}+\frac{d}{2z}-\frac{d^2}{4z^2}}{\left(1-\frac{d^2}{4z^2}\right)^2}$$

$$E = \frac{q}{4\pi\varepsilon_0 z^2}\frac{\frac{4d}{2z}}{\left(1-\frac{d^2}{4z^2}\right)^2}$$

$$E = \frac{q}{4\pi\varepsilon_0 z^2}\frac{\frac{2d}{z}}{\left(1-\frac{d^2}{4z^2}\right)^2}$$

Moving the z in the denominator of the term in the numerator of the fraction on the right down with the other z's in the denominator of the fraction on the left, and canceling the 2/4 into 1/2,

$$E = \frac{q}{2\pi\varepsilon_0 z^3} \frac{d}{\left(1 - \frac{d^2}{4z^2}\right)^2} \qquad (9\text{-}12)$$

which is our final *exact* expression. We specified that z was a long way up the z-axis, however, or, in other words, that $z \gg d$. In this case, $d/z \ll 1$, and $d^2/4z^2 \approx 0$, and we can ignore the term entirely. Now we are left with

$$E = \frac{qd}{2\pi\varepsilon_0 z^3} \qquad (9\text{-}13)$$

We define p, the "dipole moment," as qd. Our final expression for the electric field along the axis of but far away from the electric dipole is

$$E = \frac{p}{2\pi\varepsilon_0 z^3} \qquad (9\text{-}14)$$

Now that we've done one problem with a dipole, let's do a slightly harder one. Instead of looking at P along the axis that the charges are found on, let's put the charges on the x-axis and look at the electric field at some point along the y-axis midway between the two charges:

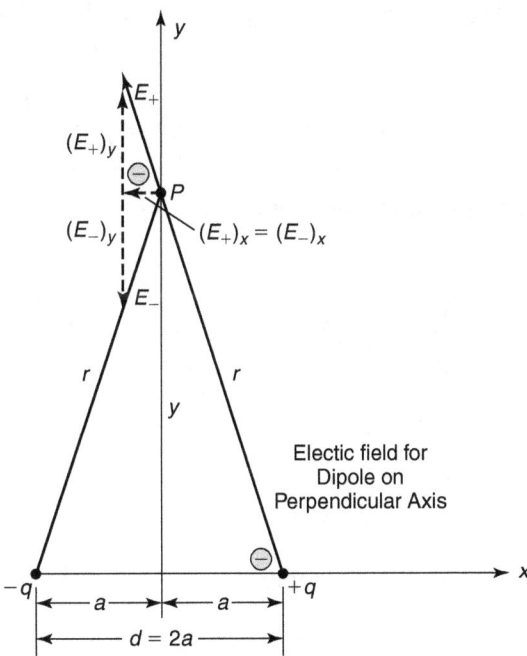

We'll specify that the two charges are again equal in magnitude and opposite in sign to each other, that P is much greater than d, and that the line from P to d bisects d.

We use Coulomb's law to write general expressions for E_1 and E_2, using the Pythagorean theorem to substitute $a^2 + y^2 = r^2$, and using a and b for the distances from each charge to the midpoint:

$$\mathbf{E}_1 = \frac{1}{4\pi\varepsilon_0} \frac{q_1}{a^2 + y^2} \mathbf{r}_1 \qquad (9\text{-}15)$$

$$\mathbf{E}_2 = \frac{1}{4\pi\varepsilon_0} \frac{q_2}{b^2 + y^2} \mathbf{r}_2 \qquad (9\text{-}16)$$

Letting θ be the angle marked and ϕ be the corresponding angle for the negative charge,

$$\mathbf{r}_1 = \cos\phi\,\mathbf{x} + \sin\phi\,\mathbf{y} \text{ and} \qquad (9\text{-}17)$$

$$\mathbf{r}_2 = \cos\theta\,\mathbf{x} + \sin\theta\,\mathbf{y} \qquad (9\text{-}18)$$

So

$$\mathbf{E}_1 = \frac{1}{4\pi\varepsilon_0}\frac{q_1}{a^2 + y^2}(\cos\phi\,\mathbf{x} + \sin\phi\,\mathbf{y}) \qquad (9\text{-}19)$$

$$\mathbf{E}_2 = \frac{1}{4\pi\varepsilon_0}\frac{q_2}{b^2 + y^2} = (\cos\theta\,\mathbf{x} + \sin\theta\,\mathbf{y}) \qquad (9\text{-}20)$$

Adding the *x*- and *y*-components together separately to find the components for the total field, $\mathbf{E}_1 + \mathbf{E}_2$,

$$E_x = \frac{1}{4\pi\varepsilon_0}\left(\frac{q_1}{a^2 + y^2}\cos\phi + \frac{q_2}{b^2 + y^2}\cos\theta\right) \qquad (9\text{-}21)$$

$$E_y = \frac{1}{4\pi\varepsilon_0}\left(\frac{q_1}{a^2 + y^2}\sin\phi + \frac{q_2}{b^2 + y^2}\sin\theta\right) \qquad (9\text{-}22)$$

Since we have a dipole, $a = b$, $q_1 = -q_2$ and $\theta = 180° - \phi$, so $\cos\theta = \cos(180° - \phi) = -\cos\phi$, while $\sin\theta = \sin(180° - \phi) = \sin\phi$. Plugging these identities into the above expressions,

$$E_x = \frac{1}{4\pi\varepsilon_0}\left(\frac{-q_2}{a^2 + y^2}\cos\phi - \frac{q_2}{a^2 + y^2}\cos\phi\right) \qquad (9\text{-}23)$$

This can be simplified by recognizing that $\cos\phi = a/r = a/\sqrt{(a^2 + y^2)}$

$$E_x = \frac{1}{4\pi\varepsilon_0}\left(\frac{-q_2}{a^2 + y^2}\frac{a}{\sqrt{a^2 + y^2}} - \frac{q_2}{a^2 + y^2}\frac{a}{\sqrt{a^2 + y^2}}\right)$$

$$E_x = \frac{1}{4\pi\varepsilon_0}\left(\frac{-2q_2}{a^2 + y^2}\frac{a}{\sqrt{a^2 + y^2}}\right)$$

$$E_x = \frac{1}{2\pi\varepsilon_0}\left(\frac{-q_2}{a^2 + y^2}\frac{a}{\sqrt{a^2 + y^2}}\right)$$

Any number times its square root is that number raised to the power of 3/2, so this can be written more simply as

$$E_x = \frac{1}{2\pi\varepsilon_0}\frac{-q_2 a}{(a^2 + y^2)^{3/2}} \qquad (9\text{-}24)$$

Let's do the same process for E_y:

$$E_y = \frac{1}{4\pi\varepsilon_0}\left(\frac{-q_2}{a^2 + y^2}\sin\phi + \frac{q_2}{a^2 + y^2}\sin\phi\right) = 0 \qquad (9\text{-}25)$$

The y-component is zero because the charges have opposite signs—the electric field lines going up from the positively charged particle cancel the electric field lines going down from the negatively charged one:

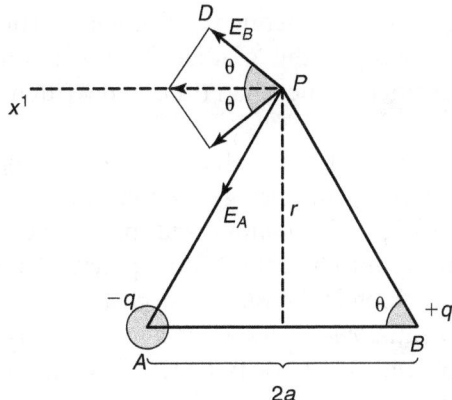

We can leave the final expression as

$$E = \frac{1}{2\pi\varepsilon_0} \frac{-qa}{(a^2 + y^2)^{3/2}} \qquad (9\text{-}26)$$

or we can define the dipole moment $p = qd = 2qa$ and write

$$E = \frac{1}{4\pi\varepsilon_0} \frac{-p}{(a^2 + y^2)^{3/2}} \qquad (9\text{-}27)$$

If $y \gg a$ as is usually the case for a dipole (the condition that the two charges are "close together"), then $y^2 \gg a^2$, and we can leave out the a^2 again:

$$E = \frac{1}{4\pi\varepsilon_0} \frac{-p}{y^3} \qquad (9\text{-}28)$$

Note that we've only proved this formula for nice symmetric geometries. We could derive a general expression for P being placed anywhere, like this:

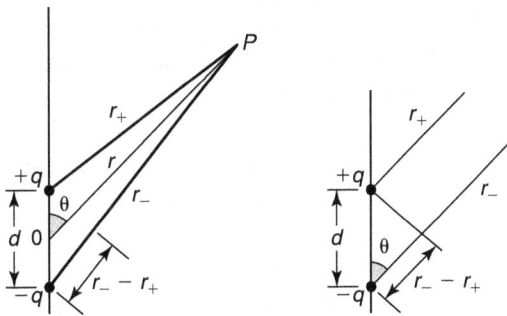

but this problem will be much easier to solve after we introduce the concept of electric potential.

Electric Potential Energy

In Chapter 6, we made the claim that for every *conservative force* independent of the path taken, every point in space could be described by a "potential" of which the force is the "steepness" or rate of change (which we defined as a simple rate of change—taking the infinitesimal limit into calculus, it becomes the "gradient," a type of three-dimensional derivative taken with respect to displacement rather than time, and written with an upside-down delta sign).

Only two potentials were introduced at the time—gravitational potential energy corresponding to the gravitational force, and spring potential energy associated with (or in philosophically looser language, "stored in") a compressed spring. The electric force is conservative as well, however, and it's time to introduce the electric potential energy.

We have to be a little more careful than we were in chapter 6 now. In chapter 6, we discussed gravitational and spring potential *energy*; now, we will be discussing both electric potential energy and something simply called "electric potential". It will be important to understand the distinction between electric potential and electric potential energy, two completely different but related, and easily confused, concepts. We will also have to be mathematically imprecise, since electric potential energy is, like any other potential, a "scalar field" in the mathematical sense employed in Chapter 6—a set of scalars associated with each coordinate point. However, our definition of the "electric field" **E**, called such because it also happens to be a field (a vector field), makes it confusing to call the potential a field as well. The mathematicians will have to grin and bear it, and the rest of us can save ourselves a lot of grief by avoiding the term *field* for anything but **E**.

For a situation in which only conservative forces (such as **F**) are present, the work-potential energy theorem tells us that

$$W_c = -\Delta U \tag{9-29}$$

Or

$$U = -W_c \tag{9-30}$$

We can employ the definition of work

$$W = F * \Delta x \tag{9-31}$$

while noting that for the electric force **F**, the displacement Δx is written with the symbol **r**. (This is because, although this has been slipped under the rug while the wool has been pulled over your eyes for the sake of simplifying the mathematical presentation, we've surreptitiously changed from "Cartesian coordinates" using rectilinear x, y, and z axes to "polar coordinates" using the *radial* displacement **r**—useful because it can point in any direction, leaving our equation intact no matter which direction it is pointing. This, of course, assumes that the electric field obeys spherical spatial symmetries, acting in the same direction without knowing east from west or up from down, a fact which is of course true. The other two "polar coordinates" specify the angular position of the displacement in three dimensions, and are not needed in Coulomb's law because of the spherical symmetries.) Rewriting the work using Coulomb's law for the force and **r** for Δx,

$$W = F * \Delta x = F * r = k_e \frac{q_1 q_2}{r^2} * r = k_e \frac{q_1 q_2}{r} = -U \tag{9-32}$$

Or

$$U = -k_e \frac{q_1 q_2}{r} \tag{9-33}$$

As mentioned previously in the text, for electricity it is convenient to make a distinction between the electric potential energy itself (U), which is a property *between* two things given any separation between them r, and the "electric potential" called V, which is a property *just* of the space, or more precisely a scalar field. The relationship between the two is simple: V is the potential energy divided by the charge that otherwise would have occupied the place at which V is being measured:

$$V \equiv \frac{U}{q_2} = k_e \frac{q}{r} \tag{9-34}$$

This is mathematically a scalar field and is very analogous to our definition of the vector field, which we are calling "electric field":

$$\mathbf{E} \equiv \frac{\mathbf{F}}{q_2} \tag{9-35}$$

For the sake of convenience, let's list all four quantities we have defined in a table:

$$\mathbf{F} = k_e \frac{q_1 q_2}{r^2} \hat{r} \qquad \mathbf{E} = k_e \frac{q}{r^2} \hat{r} \tag{9-36}$$

$$U = k_e \frac{q_1 q_2}{r} \qquad V = k_e \frac{q}{r}$$

The two vector quantities and the two scalar quantities are related to each other by the multiplication of a charge:

$$\mathbf{F} = q\mathbf{E} \tag{9-37}$$

$$U = qV \tag{9-38}$$

Likewise, the vector quantities are related to their scalar counterparts going vertically in the table by the dot product between the vector quantity and the displacement r (not to be confused with the unit vector $\hat{r}$):

$$U = \mathbf{F} * \mathbf{r} \tag{9-39}$$

$$V = \mathbf{E} * \mathbf{r} \tag{9-40}$$

These relationships reveal some interesting facts about units. We already knew from our definitions in chapter 6 that a Joule (the unit of energy) was a Newton-meter (N *m); equation 9-38 tells us that a Joule is also a Coulomb-Volt. The unit of electric field can be given as either Volts per meter or as Newtons per Coulomb.

Electric Field Lines

The work done to move a particle between two points using the electric force is given by the *change* in electric potential energy—just like the gravitational potential energy, only the change has real physical meaning, not the "zero point" that we've conveniently assigned in the definitions (infinite distance apart for the potential energy between two particles, and equilibrium position for a spring). A similar definition is taken for the electric potential V from a point charge. However, most of the advanced problems involve not simple ideal point charges, but various distributions of electric field due to a macroscopic number of charges, for example, in a circuit. In this case, it is simply convenient to pick a "ground" where the electric potential—identical to what in a circuit is called "voltage"—is declared to be 0. Only the potential difference matters and has physical meaning.

Electric fields are often represented by electric field lines. These are lines that graphically depict the magnitude of the electric field lines through the following simple rules:

1. Electric field lines always come out of the positive charge or charge distribution, and go into the negative charge distribution.
2. The density of electric field lines is proportional to the strength of the electric field.

The field lines map the direction of the path of a charged particle in the presence of that electric field. Charged particles "fall" to a lower potential ("downhill," to take an analogy with gravitational potential) along the direction of the electric field lines, and unlike anything ever seen with the gravitational force, negatively charged particles move in opposite direction of the electric field lines, falling "uphill." Perpendicular to these electric field ones one can draw

228 Foundations & Principles of Physics

equipotential lines, graphs of lines in space where the electric potential is equal. Equipotential lines can be thought of as "contour lines" mapping out the density of the electric field in the same way that altitude lines mark out elevation on a contour map. Both are graphs showing where in space the potential (gravitational potential in the latter case, electric potential in the former) is equal. The closer the lines are together, the steeper the slope (which is mathematically called the gradient) is, and the more work must done on an object to move it between those two points; a more advanced mathematical treatment would define the electric field as being nothing other than the gradient or slope of the potential.

There are a number of reasons for preferring to work with U and V rather than with $\mathbf{F}$ and $\mathbf{E}$, although it means defining two more new terms. V is a scalar and only has r rather than r^2 in the denominator, making it much simpler to work with. r is the magnitude of the displacement, the distance from a charge to the point, without containing information about direction or having any vector components to be separated out and manipulated individually. Like $\mathbf{F}$ and $\mathbf{E}$, it obeys the principle of superposition—the electric potential at a point from three particles is the potential from each particle added together—but its scalar nature makes it much simpler to add. We can illustrate the computational power it gives us by comparing it with what for electric field was a particularly difficult example: let's find the electric *potential* from a dipole, measured at **any** point P, not just carefully chosen Ps where the symmetries make it easy (as we did when calculating $\mathbf{E}$ at P).

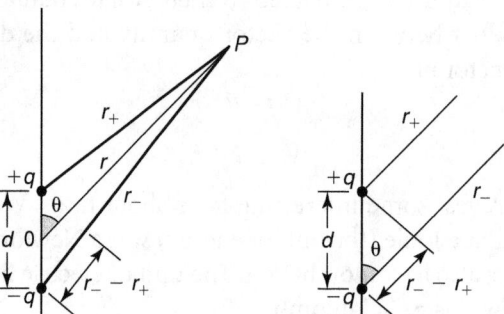

Since electric potential obeys the principle of superposition,

$$V = V_+ + V_- = \frac{1\left(\dfrac{q}{r_+} + \dfrac{-q}{r_-}\right)}{(4\pi\varepsilon_0)} \tag{9-41}$$

Putting the two terms over a common denominator,

$$V = \frac{q}{4\pi\varepsilon_0}\frac{r_- - r_+}{r_-r_+} \tag{9-42}$$

At this point, dealing with as general terms as we are, we can go no further. However, if $r \gg d$ as is the case for a dipole, then r_+ and r_- are practically parallel to each other, as is shown in the diagram on the right, and from geometry we can see that $r_- - r_+ \approx d\cos\theta$. Also, since r_+ and r_- and r are all about the same length relative to d, we can say that $r_-r_+ \approx r^2$. Plugging these approximations in,

$$V = \frac{q}{4\pi\varepsilon_0}\frac{d\cos\theta}{r^2} \tag{9-43}$$

Using $p = qd$ again,

$$V = \frac{p\cos\theta}{4\pi\varepsilon_0 r^2} \tag{9-44}$$

Since $V = \mathbf{E} * \mathbf{r}$, $E = V/r$, so for a dipole with any arbitrary point P,

$$E = \frac{p\cos\theta}{4\pi\varepsilon_0 r^3} \tag{9-45}$$

HOMEWORK FOR CHAPTER 9

Name _____

A. You discover an old treasure map that gives the directions to buried treasure. Unfortunately, in order to throw off the villainous but dim-witted enemies, the pirate had written the location as a vector sum. The instructions read, "Matey, go east 30 paces. Drink a bottle o' rum and go 40 paces at a 27° angle counterclockwise from east. Go 25 paces rotating 2.5 radians from east. Go 12 paces due south and then 50 paces rotating 1.8 radians from east." (Must be read in a pirate accent in order to get the correct directions.) Draw a map of your path and give the coordinates of the buried treasure, if you start at (0,0).

B. Add $(7, 13) + (5, -8)$.

C. Subtract the vectors $(-12, 0) - (4, 7)$.

D. Take the cross product between the vectors $(5, 6)$ and $(-1, 8)$.

E. Graph the following vectors.
   1. Graph the vector (8, 12).

2. Graph the vector (0, 5).

3. Graph the vector $(-5, 7)$.

4. Graph the vector (2, 2).

5. Graph the vector 7 @ 38°.

6. Graph the vector 9 @ $\pi/6$.

7. Graph the vector 4 @ $3\pi/2$.

8. Graph the vector $-5 \ @ \ 3\pi/4$.

F. A vector **A** has magnitude 25 at an angle 25°. Find the *x*- and *y*-components of **A**.

G. A vector **B** has components 17**x** + 18**y**. Find the magnitude and direction of **B**.

H. A vector **C** has magnitude $\sqrt{2}$ at an angle 45°. Find the *x*- and *y*-components of **C**.

I. A vector **D** has components 5**x** + 5$\sqrt{3}$**y**. Find the direction and magnitude of **D**.

J. A vector **A** has magnitude 25 at an angle 25°, and a vector **C** has magnitude $\sqrt{2}$ at an angle 45°. Find the magnitude and direction of **A** + **C**.

K. A vector $\mathbf{E_1}$ has magnitude $1.3 * 10^6$ at an angle $74°$, and a vector $\mathbf{E_2}$ has magnitude $8.2 * 10^5$ at an angle $106°$. Find the magnitude and direction of $\mathbf{E_1} + \mathbf{E_2}$.

L. A vector $\mathbf{F_1}$ has magnitude 580 at an angle $70°$, and a vector $\mathbf{F_2}$ has magnitude -100 at an angle $90°$. Find the magnitude and direction of $\mathbf{F_1} + \mathbf{F_2}$.

M. A point $\mathbf{A}$ exists on a Cartesian coordinate system at location $(-2, 0)$, and a point $\mathbf{B}$ exists on a Cartesian coordinate system at location $(0, 7)$. Find the magnitude and direction of the distance $\mathbf{r}$ between $\mathbf{A}$ and $\mathbf{B}$.

N. A point $\mathbf{A}$ exists on a Cartesian coordinate system at location $(-2, 0)$, a point $\mathbf{B}$ exists on a Cartesian coordinate system at location $(0, 7)$, and a point $\mathbf{C}$ exists on a Cartesian coordinate system at location $(3, 0)$. Let $\mathbf{r_1}$ be the distance from $\mathbf{A}$ to $\mathbf{B}$ and $\mathbf{r_2}$ be the distance from $\mathbf{B}$ to $\mathbf{C}$. Find the magnitude and direction of $\mathbf{r_1} + \mathbf{r_2}$.

O. A point charge q_1 with magnitude 5 C exists on a Cartesian coordinate system at location $(-2, 0)$, a point charge q_2 with magnitude $-5C$ exists on a Cartesian coordinate system at location $(0, 7)$, and a point charge q_3 with magnitude 5 C exists on a Cartesian coordinate system at location $(3, 0)$. Let $\mathbf{r}_1$ be the distance from q_1 to q_2, and $\mathbf{r}_2$ be the distance from q_3 to q_2. Let $\mathbf{F}_1$ be the electric force between q_1 and q_2 and $\mathbf{F}_2$ be the electric force between q_3 and q_2. Find the magnitude and direction of $\mathbf{F}_1 + \mathbf{F}_2$.

Note: Coulomb's law tells us that $\mathbf{F}_1 = (8.99 * 10^9) \dfrac{q_1 q_2}{\mathbf{r}_1^2}$ and $\mathbf{F}_2 = (8.99 * 10^9) \dfrac{q_3 q_2}{\mathbf{r}_2^2}$

P. A particle Q_1 at the origin has a charge 5 C and a particle Q_2 at $x = 2$ has a charge $+3C$. Find the force on the particle Q_3 at x = 4, which has a charge 2 C.

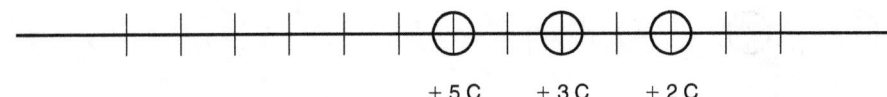

Q. Take the same situation as in problem A, but let the particle at x = 2 have a charge of −3C. What is the force on the particle Q_3 at x = 4, with a charge +2 C?

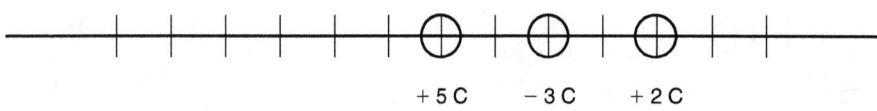

R. If the particle Q_3 at x = 4 in problem Q had a charge of 0 C instead of +2C, what would the force on it be?

S. Suppose you have a 5 nC charge at the origin and a 2 nC charge at $x = 5$ m. Find the location on the x axis where the electric field is 0. (Find a finite coordinate location.)

T. Suppose you again have a 5 nC charge at the origin and a 2 nC charge at $x = 5$ m. Find (a) the electric field at $x = -1$ m, and (b) the electric force (magnitude and direction) on a −6 nC charge placed at $x = -1$ m.

U. Suppose you have a 3 C charge at $x = 1$ m and a -3 C charge at $x = -1$ m. Find (a) the electric field at $x = 3$ m, and (b) the electric force on a charge of 3 nC placed at $x = 3$ m. Careful with units.

V. Suppose you have a 3 C charge at $x = 1$ m and a -3 C charge at $x = -1$ m. Without using the dipole approximation, find (a) the electric field at $x = 25$ m, and (b) the electric force on a charge of 75 nC placed at $x = 25$ m. Show your work.

W. Suppose you have a 3 C charge at $x = 1$ m and a -3 C charge at $x = -1$ m. Using the dipole approximation, find (a) the electric field at $x = 25$ m, and (b) the electric force on a charge of 75 nC placed at $x = 25$ m. Show your work.

X. Suppose you have a 3 C charge at (1, 0) m and a −3 C charge at (−1, 0) m. Find (a) the electric field at the point (0, 3), and (b) the electric force on a charge of 3 nC placed at (0,3). Show all your work.

Y. Suppose you have a 3 C charge at (1, 0) m and a −3 C charge at (−1, 0) m. Without using a dipole approximation, find (a) the electric field at the point (0, 50), and (b) the electric force on a charge of 3 nC placed at (0, 50). Show all your work.

Z. Suppose you have a 3 C charge at (1, 0) m and a −3 C charge at (−1, 0) m. Using a dipole approximation, find (a) the electric field at the point (0, 50), and (b) the electric force on a charge of 3 nC placed at (0, 50). Show all your work.

AA. A particle Q_1 at $x = -1$ has a charge $+5C$, and another particle Q_2 at $x = +1$ has a charge $-5C$. Find the electric field at point P, which is at $x = 16$?

(*Hint:* The electric field from a dipole is $E = \dfrac{1}{2\pi\varepsilon_0} \dfrac{p}{r^3}$)

BB. Find the potential energy of a system of two point charges shown in the figure.

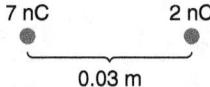

CC. Find the potential energy of a system of two point charges shown in the figure. Note that the figure is not the same as in problem A.

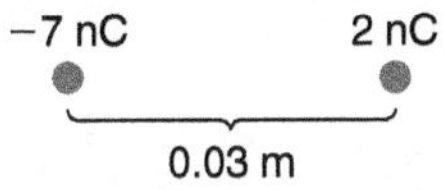

−7 nC   2 nC
  •────────•
      0.03 m

DD. Find the work that should be done to increase the distance between two point charges with the charges of 5 nC and −10 nC by 5 cm. The initial distance between the charges is 15 cm.

EE. Find the point at which the electric potential due to two point charges separated by a distance of 4 m is 0.

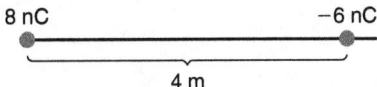

FF. Find the electric potential and the electric field at point P due to a system of three point charges as shown in the figure.

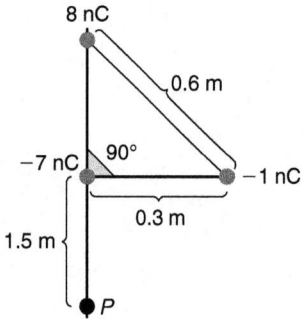

GG. In your parents' generation, televisions used devices called "cathode ray tubes" in which the images on the screen were generated by electrons fired in a vacuum from the back of the television to the screen, across a potential difference of about 25 kV. If the distance from the back of the screen to the front was 6 feet (not too unreasonable for an old clunker, give or take an order of magnitude or two), find the magnitude of the uniform electric field between the front and back of the television across which the electrons would accelerate. Hint: Remember to convert feet to SI units.

HH. You've heard that physics research is a fast way to get rich, and you want to make it big in the physics world. You want to impress your professor because he is clearly just rolling in dough, so you write a paper that will knock his socks off explaining why the helium nucleus is stable. You know that the proton has the same charge as the electron except that the proton charge is positive. Neutrons you know are neutral. Why, your paper asks, don't the protons simply repel each other causing the helium nucleus to fly apart? Obviously we need to start thinking outside the box. You remember from Chapter 6 that we talked about another conservative force besides the electric force, namely the gravitational force. So the model presented by your paper suggests that the two neutrons sit in the center of the nucleus and gravitationally attract the two protons. Since the protons have the same charge, they are always as far apart as possible on opposite sides of the neutrons. But before you show your paper to your professor, you'll want to check the model with a few calculations. What mass would the neutron have if this model of the helium nucleus works? Is that a reasonable mass? (Recall that the mass of a neutron is about the same as the mass of a proton and that the diameter of a helium nucleus is 3.0×10^{-15} m.)

II. Take a point charge +5nC at x = −1 and a point charge +5nC at x = 1. We will use the following steps to find the electric field and potential at some point P on the y-axis at y = 3:

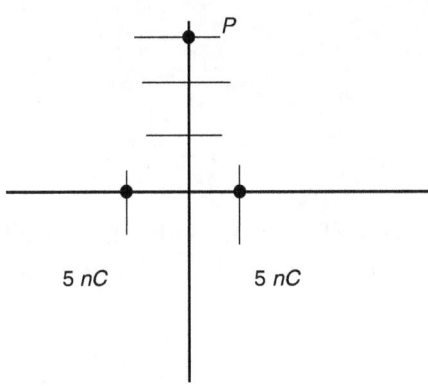

1. Find the length of the distance r_1 from the 5 nC charge at x = −1 to the point P. Use the Pythagorean theorem or the distance formula, whichever you feel more comfortable with.
   Hint:

$$r_1 = \sqrt{x^2 + y^2} \tag{9-46}$$

2. Find the length of the distance r_2 from the 5 nC charge at x = 1 to the point P. Use the Pythagorean theorem or the distance formula.

3. Find the contribution V_1 to the electric potential at P from the 5 nC charge at x = −1.
   Hint:

$$V = k_e \frac{q}{r_1} \tag{9-47}$$

4. Find the contribution V_2 to the electric potential at P from the 5 nC charge at x = 1.

5. Find the total electric potential V = V$_1$ + V$_2$. *Hint:* These are scalars; you just need to add them together.

6. Find the magnitude **and** direction of the electric field contribution **E**$_1$ at point P caused by just the charge at $x = -1$. (In other words, find what the electric field at P would be if the 5 nC charge at $x = -1$ were the only charge present.)
   Hint: $\mathbf{E}_1 = \dfrac{q}{r_1^2}$ and $\theta = \tan^{-1}\left(\dfrac{y}{x}\right)$

7. Find the magnitude **and** direction of the electric field contribution **E**$_2$ at point P caused by just the charge at $x = 1$. (In other words, find what the electric field at P would be if the 5 nC charge at $x = 1$ were the only charge present.)

8. Find the x-component of **E**$_1$.

9. Find the *y*-component of **E₁**.

10. Find the *x*-component of **E₂**.

11. Find the *y*-component of **E₂**.

12. Find **Eₓ**, the *x*-component of the **total** electric field at P.

13. Find **E_y**, the *y*-component of the **total** electric field at P.

14. Find the magnitude **and** direction of the total electric field at P.

JJ. Take a point charge +5 nC at x = −1 and a point charge +5 nC at x = 2. We will use the following steps to find the electric field and potential at some point P on the *y*-axis at $y = 3$:

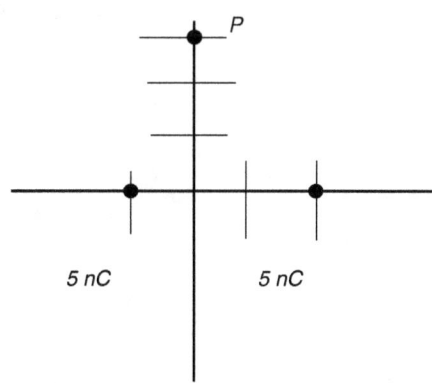

1. Find the length of the distance r_1 from the 5 nC charge at $x = -1$ to the point P. Use the Pythagorean theorem or the distance formula, whichever you feel more comfortable with.
   Hint: $r_1 = \sqrt{x^2 + y^2}$

2. Find the length of the distance r_2 from the 5 nC charge at $x = 2$ to the point P. Use the Pythagorean theorem or the distance formula.

3. Find the contribution V_1 to the electric potential at P from the 5 nC charge at $x = -1$.
   Hint: $V = k_e \dfrac{q}{r_1}$

4. Find the contribution V_2 to the electric potential at P from the 5 nC charge at $x = 2$.

5. Find the total electric potential $V = V_1 + V_2$. *Hint:* These are scalars; you just need to add them together.

6. Find the magnitude **and** direction of the electric field contribution $\mathbf{E}_1$ at point P caused by just the charge at $x = -1$. (In other words, find what the electric field at P would be if the 5 nC charge at $x = -1$ were the only charge present.)
   Hint: $\mathbf{E}_1 = k_e \dfrac{q}{r_1^2}$ and $\theta = \tan^{-1}\left(\dfrac{y}{x}\right)$

7. Find the magnitude **and** direction of the electric field contribution E_2 at point P caused by just the charge at $x = 2$. (In other words, find what the electric field at P would be if the 5 nC charge at $x = 2$ were the only charge present.)

8. Find the x-component of $\mathbf{E}_1$.

9. Find the y-component of $\mathbf{E}_1$.

10. Find the x-component of $\mathbf{E}_2$.

11. Find the y-component of $\mathbf{E}_2$.

12. Find $\mathbf{E}_x$, the x-component of the **total** electric field at P.

13. Find $\mathbf{E}_y$, the y-component of the **total** electric field at P.

14. Find the magnitude **and** direction of the total electric field at P.

266 Foundations & Principles of Physics

KK. Take a point charge +5 nC at x = −1 and a point charge +7 nC at x = 2. We will use the following steps to find the electric field and potential at some point P on the y-axis at y = 3:

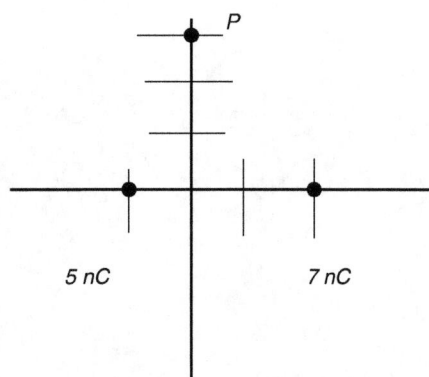

1. Find the length of the distance r_1 from the 5 nC charge at x = −1 to the point P. Use the Pythagorean theorem or the distance formula, whichever you feel more comfortable with.
   Hint: $r_1 = \sqrt{x^2 + y^2}$

2. Find the length of the distance r_2 from the 7 nC charge at x = 2 to the point P. Use the Pythagorean theorem or the distance formula.

3. Find the contribution V_1 to the electric potential at P from the 5 nC charge at x = −1.
   Hint: $V = k_e \dfrac{q}{r_1}$

4. Find the contribution V_2 to the electric potential at P from the 7 nC charge at x = 2.

CHAPTER 9 Coulomb's Law **267**

5. Find the total electric potential V = V$_1$ + V$_2$. *Hint:* These are scalars; you just need to add them together.

6. Find the magnitude **and** direction of the electric field contribution **E**$_1$ at point P caused by just the charge at $x = -1$. (In other words, find what the electric field at P would be if the 5 nC charge at $x = -1$ were the only charge present.)
   Hint: $\mathbf{E}_1 = k_e \dfrac{q}{r_1^2}$ and $\theta = \tan^{-1}\left(\dfrac{y}{x}\right)$

7. Find the magnitude **and** direction of the electric field contribution **E**$_2$ at point P caused by just the charge at $x = 2$. (In other words, find what the electric field at P would be if the 7 nC charge at $x = 2$ were the only charge present.)

8. Find the *x*-component of **E**$_1$.

9. Find the *y*-component of **E**$_1$.

10. Find the *x*-component of **E**$_2$.

11. Find the *y*-component of **E**$_2$.

12. Find **E**$_x$, the *x*-component of the **total** electric field at P.

13. Find **E**_y_, the *y*-component of the **total** electric field at P.

14. Find the magnitude **and** direction of the total electric field at P.

15. Find the total electrical potential energy between the 5 nC and 7 nC charges.
    Hint:
    $$\Delta U = k_e \frac{q_1 q_2}{r} \qquad (9\text{-}48)$$

LL. Take a point charge +5 nC at x = −1 and a point charge +7 nC at x = 2. Now there is a point charge +4 nC on the *y*-axis at y = 3:

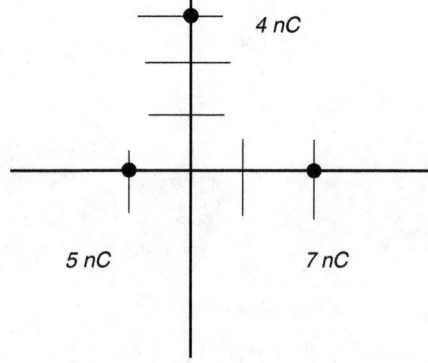

1. Find the length of the distance r_1 between the 5 nC charge and the 4 nC charge. Use the Pythagorean theorem or the distance formula, whichever you feel more comfortable with.
   Hint: $r_1 = \sqrt{x^2 + y^2}$

2. Find the length of the distance r_2 between the 7 nC charge and the 4 nC charge. Use the Pythagorean theorem or the distance formula.

3. Find the electrical potential energy U_1 between the 5 nC charge and the 4 nC charge.
   Hint: $\Delta U = k_e \dfrac{q_1 q_2}{r}$

4. Find the electrical potential energy U_2 between the 7 nC charge and the 4 nC charge.

5. Find the total electrical potential energy between **all three** charges. *Hint:* This is a sum of three terms, each of which has the form of the hint in part 3 of this problem.

6. Find the magnitude **and** direction of the electric force F_1 between the 5 nC and 4 nC charges.
   Hint: $F_1 = k_e \dfrac{q_1 q_2}{r_1^2}$ and $\theta = \tan^{-1}\left(\dfrac{y}{x}\right)$

7. Find the magnitude **and** direction of the electric force F_2 between the 7 nC and 4 nC charges.

8. Find the *x*-component of F_1.

9. Find the *y*-component of F_1.

10. Find the *x*-component of F_2.

11. Find the *y*-component of **F**$_2$.

12. Find **F**$_x$, the *x*-component of the **total** electric force on the 4 nC charge. *Hint*: We don't care about the force between the 5 nC and 7 nC charges in this problem.

13. Find **F**$_y$, the *y*-component of the **total** electric force on the 4 nC charge.

14. Find the magnitude **and** direction of the total electric force on the 4 nC charge.

Gauss's Law

10

Flux and Gauss's Law

In Chapter 9, we found ways to calculate the electric field due to single point charges and collections of small numbers of point charges. In the real world, however, most charge comes from *continuous charge distributions*, charged objects that contain anywhere from billions to 10^{25} charged particles. Trying to use Coulomb's Law to find the electric field from each single individual charged particle—while worrying about the fact that subatomic particles are always in motion—is simply prohibitively complicated. Instead, we'll have to find large-scale, macroscopic ways of thinking about electric field, and we'll derive a simple expression—named after Carl Friedrich Gauss, who formulated it in 1835—called Gauss' Law. Gauss represented electric field by "electric field lines"—lines of force indicating the direction of motion of positively charged particles placed in an electric field—and argued that the total number of electric field lines passing through a surface must be the density of those field lines times the area of that surface. The key idea allowing us to perform calculations is that the electric field **E** discussed in Chapter 9 is nothing other than a measurement of the *density* of electric field lines. This is actually little more than a definition; electric field lines are a tool devised to represent electric fields, with the electric stronger the closer the lines are spaced, and the electric field being weaker the farther apart they are. Electric field lines are defined so that **E** represents their density. However, definitions are made because they are fruitful, and this definition is no exception.

The total number of electric field lines passing through an area is called *electric flux*, and itís represented by the symbol Φ. Since the electric field **E** is the density of electric field lines, the total number of electric field lines passing through a surface is the density **E** times their area **A**. In mathematical terms,

$$\Phi = \mathbf{E} * \mathbf{A}$$

You might be surprised to see the letter A put in bold as a vector. Area isn't something we normally think of as a vector. However, the electric flux Φ is a scalar—it's simply the number of electric field lines passing through a surface, nothing more, and simple numbers aren't vectors. If Φ is a scalar, then whatever goes on the other side of the equal sign must also be a scalar—nothing can be both a scalar and a vector at the same time, so scalars can only be equal to scalars, and vectors to vectors We could have given the magnitude of *E* instead of the full vector **E** on the other side, but it is pretty important which direction the electric field is pointing, since the field lines point in the direction of the electric field. If the electric field lines all run along the edge parallel to the surface A, none of them will cross through the surface and the electric flux will be 0, not EA. The electric flux will be maximized as EA when the field lines are perpendicular to the surface, allowing the maximum number of them to cross through it. So we need the right-hand side of the equation to be a function of vectors that gives us a scalar. The dot product just so happens to be such an operator, but it requires that **A** be a vector as well.

The dot product **E*A** equals $EA \cos \theta$. We want the dot product to be maximized when the electric field lines are perpendicular to the surface, but $\cos \theta$ is maximized when $\theta = 0$. Consequently, we need the vector **A** to also point perpendicular to the surface, so that the angle between it and **E** is 0 when the flux is maximized. **A** is therefore a vector, with magnitude equal to the area of the surface, pointing perpendicular to (or normal to) that surface.

Let's derive Gauss's law by finding the flux through a spherical surface of radius r enclosing a point charge q. We know from Coulomb's law that the electrical field at any point on that surface (since they all have the same distance from the charge, r) is $\frac{1}{4\pi\epsilon_0} \frac{q}{r^2}$. Likewise, we know that the surface area of a sphere is $4\pi r^2$. So $\Phi = \frac{1}{4\pi\epsilon_0} \frac{q}{r^2} * 4\pi r^2 = \frac{q}{\epsilon_0}$. Although it takes calculus to prove this, it turns out that for *any* distribution of charges with *any* surface perpendicular to all of the electric field lines in such a way that the electric field is uniform around the surface, $EA = \frac{q_{enc}}{\epsilon_0}$. This result is called **Gauss's law**. (One can express Gauss's law, as Gauss himself did, without the conditions of uniformity and perpendicularity, but doing so requires calculus, and is usually not solvable.)

We use Gauss's law by finding an appropriate imaginary surface to find the flux going through (called a "Gaussian surface"), and then substituting EA for the flux, then dividing by A to get the electric field at every point around that surface. Let's do two examples.

Example 10-1. Take a long, nonconducting wire with a uniform charge density of λ C/m. (One-dimensional have a charge density in C/m, and the symbol is usually λ.) Find the electric field a distance r away from the wire.

The electric field is going to be pointing radially outward from the wire in all directions, so we need to enclose it with a surface perpendicular to a fan-shaped distribution of electric field lines. The shape we need is a cylinder with the wire going down its axis.

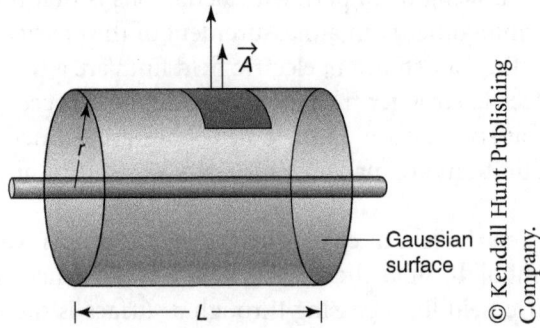

It doesn't matter how long we make the cylinder; let's say it has a length L. The field lines will only cross through the side of the cylinder, not the top and bottom. So the surface area we need is the surface area of the side of the cylinder, which is the circumference of the circle ($2\pi r$) times the length L. The total amount of charge enclosed will be the charge density λ times the length of the wire enclosed, or λL. So

$$E * 2\pi rL = \lambda L / \epsilon_0$$

Solving for E,

$$E = \frac{\lambda}{2\pi r \epsilon_0} \tag{10-1}$$

Example 10-2. Now let's use Gauss's law to find the electric field from one other surface, an infinite nonconducting sheet with uniform charge density. Let's say it has a charge density σ N/m² (σ is the standard symbol for charge density of a sheet).

If a uniformly charged sheet is big enough, the field lines will all be going straight up and down, and we need any Gaussian surface with sides parallel to the sheet. We could use a cube, or we could use a cylinder whose z axis is oriented parallel with the electric field lines, like this:

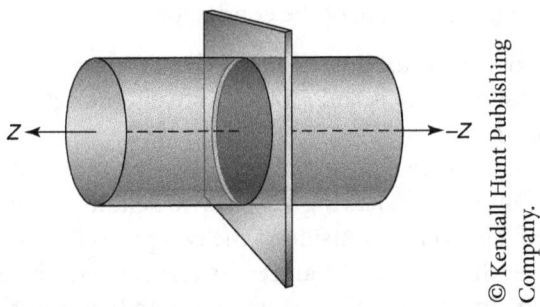

Notice that the electric field lines pass through *both* the top and the bottom of the cylinder. So the area through which we calculate the flux is $\pi s^2 + \pi s^2 = 2\pi s^2$, where s is the radius of the cylinder (it doesn't matter what value you choose for s). The total charge enclosed will be the charge density σ multiplied by the area of the sheet enclosed by the charge, which is just πs^2. So, plugging these into Gauss's law,

$$E * 2\pi s^2 = \frac{\sigma \pi s^2}{\varepsilon_0} \tag{10-2}$$

$E = \frac{\sigma}{2\varepsilon_0}$. The infinite line and infinite sheet of charge are the principal geometries you will encounter and have to use Gauss's law for.

Conductors

A conductor is an object (usually made out of metal) in which electrons are free or semifree to move around macroscopic energy states within the entire substance rather than being bound to individual molecules. As a result, they have some unique properties:

1. Any excess charge will congregate around the surface, since that is how it can get as far away as possible from the other charges.
2. Using Gauss's law to find the electric field inside of a conductor, since all the excess charge is in the middle, $q_{enc} = 0$ everywhere inside the conductor, so E = 0 everywhere inside a conductor.
3. The field outside of a conductor is always perpendicular to the surface.
4. Charge accumulates at sharp points on the surface (why you never want to stick a fork in a microwave—electrons accumulate on the prongs, and sparks start to fly).
5. Charge can be *induced* in a conductor. Bringing a negatively charged object near a conductor will repel the electrons in the conductor so that there is an imbalance in the electron distribution, with more electrons on the side opposite the negatively charged object. Attaching a grounding wire to the conductor will allow the electrons to get even further away by flowing into the ground, leaving the conductor with an overall positive charge. Likewise, bringing a positively charged object near a conductor will attract electrons, causing the side of the conductor near the positively charged object with a

slight negative charge and the opposite side with a slight positive charge. Attaching a grounding wire to the conductor will allow the positively charged side to attract electrons from the ground, canceling out that positive charge, and leaving the conductor with an overall negative charge.

6. A conductor is attracted by an electrically charged object of either charge. An electrically charged object induces the opposite charge on the side of the conductor close to the object while inducing the same charge on the opposite side (if no grounding wire is attached). Because force is inversely proportional to distance, the attractive force between the object and the near side of the conductor will be stronger than the repulsive force between the object and the far side of the conductor.

Let's relate conductors back to Gauss's law. What if we have two conducting charged plates, of equal and opposite charge σ and $-\sigma$ respectively, brought close to each other (e.g., a "parallel-plate capacitor")?

The different charges on the two plates will attract each other, and will all be pulled over to the edge adjacent to the other plate, leaving the electric field in the middle of the plates 0 and consequently the electric field on the outside of the two plates 0. Now the charge on the sides of the two plates facing each other are 2σ and -2σ respectively, since each side of each plate had a surface charge of σ, and both of them were moved to the same side. So using Gauss's law to find the electric field between the two plates, the charge density is 2σ rather than σ, and so the field is $E = \sigma/\varepsilon_0$ rather than $\sigma/2\varepsilon_0$.

Capacitance

A device constructed out of two charged plates with equal and opposite charges is called a *capacitor*, and is used for a variety of applications in electrical engineering to store and discharge small amounts of charge over short but noninfinitesimal time frames. They charge and discharge at predictable rates, making them useful for timekeeping purposes. Capacitors are characterized by a new property of the object called the *capacitance*, defined as

$$C \equiv \frac{Q}{\Delta V} \qquad (10\text{-}3)$$

where Q is the charge stored on the positive plate and ΔV is the potential difference between the plates. The capacitance is usually printed on the bottom of a capacitor, so if one knows the charge on one of the plates or the potential difference between them, the capacitance is a facile way to obtain the other value as well.

When a capacitor is made out of rectangular plates, called a "parallel-plate capacitor," a more detailed expression for the capacitance may be found. The charge on one plate is the charge density σ times the area of the plate, A, since surface charge density is defined as total charge divided by area; the potential V is a concept to be defined in the next chapter. As we'll see, for a uniform electric field $V = \mathbf{E} * \mathbf{r}$ where $\mathbf{r}$ is the distance from the source of the charge. So the potential difference ΔV between the two plates is $\mathbf{E} * \mathbf{d}$ where $\mathbf{d}$ is the distance between the two plates. Plugging these values in, we see that

$$C = \frac{\sigma A}{Ed}$$

Using $E = \dfrac{\sigma}{\epsilon_0}$ as found earlier,

$$C = \dfrac{\sigma A}{\dfrac{\sigma}{\epsilon_0} d}$$

$$C = \epsilon_0 \dfrac{A}{d} \qquad (10\text{-}4)$$

Capacitors provide a useful and easy situation for using Gauss's law to evaluate electric fields and other properties in more complicated physical systems.

HOMEWORK FOR CHAPTER 10

Name _____

A. A long, insulating wire has a uniform line charge density $\lambda = 7$ nC/m.

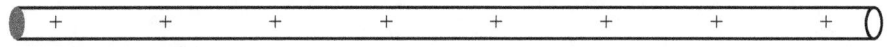

1. Draw the electric field lines around the wire.

2. Draw the Gaussian surface around the wire for a distance $r = 3$ m away from the wire. *Hint:* It must be perpendicular to *all* the electric field lines at $r = 3$ m away.

3. Find the surface area of the side of the Gaussian surface through which electric field lines are passing.

4. Find the total charge enclosed by the Gaussian surface.

5. Use Gauss's law to find the electric field at $r = 3$ m.

B. An infinite, insulating sheet has a surface charge density π C/m^2.

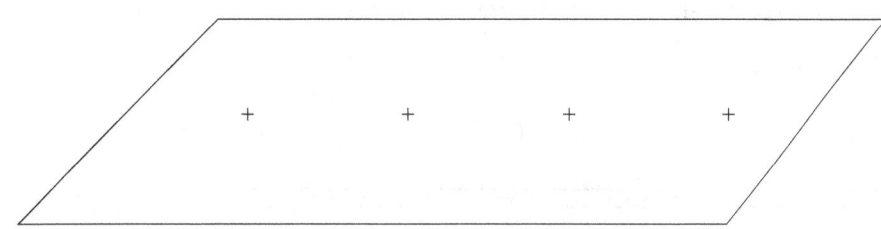

278 Foundations & Principles of Physics

1. Draw the electric field lines above and below the sheet.

2. Draw a Gaussian surface for the electric field a distance $r = 12$ km above and below the sheet.

3. Find the surface area of the Gaussian surface through which electric fields are passing.

4. Find the total charge enclosed by the Gaussian surface.

5. Use Gauss's law to find the electric field a distance $r = 12$ km *above* the conducting sheet.

C. Use Gauss's law to find the electric field at a point $r = 5$ m away from a charge $q = 7$ C. Show your work.

D. Use Gauss's law to find the electric field at a point $r = 5$ m away from an infinitely long wire with a uniform charge density $\lambda = 3$ C/m. Show your work.

++

E. Use Gauss's law to find the electric field a distance $r = 5$ m above an infinite sheet with a uniform surface charge density $\sigma = 6$ C/m².

F. A parallel-plate capacitor has rectangular plates with length 5 cm and width 3 cm. They are separated by a distance 12 m, and hooked up to a 20-V battery.
   1. What is the area of each plate, in m²?

   2. Assuming nothing else is present on the circuit, what is the potential difference between the plates? *Hint:* This one is a freebee.

   3. Recall that for a parallel-plate capacitor the capacitance is $C = \varepsilon_0 \dfrac{A}{d}$. What is the capacitance of this parallel-plate capacitor?

   4. Recall that the definition of capacitance is $C = \dfrac{Q_{tot}}{\Delta V}$. What is the total charge on the positively charged plate of this capacitor?

   5. Recall that we define surface charge density as $\sigma = \dfrac{Q_{tot}}{A}$. What is the surface charge density σ of this capacitor?

6. What is the electric field outside the capacitor?

7. Draw the electric field lines inside the capacitor in the figure below:

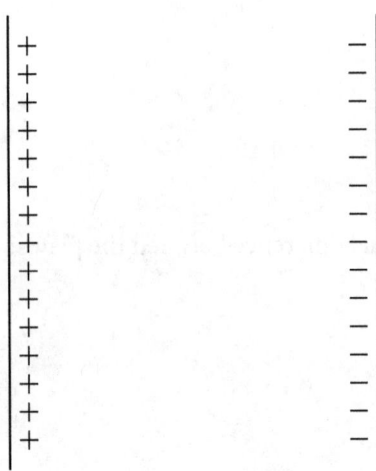

8. Draw the Gaussian surface through the positively charged plate of the capacitor. *Hint:* It must be perpendicular to *all* of the electric field lines.

9. What is the area of the Gaussian surface you drew through which electric field lines cross?

10. What is the *total* charge enclosed by the Gaussian surface you drew?

11. Using Gauss's law, what is the electric field at some point inside the capacitor?

Current and Electric Circuits

Current and Drift Velocity

In Chapters 9 and 10, we quietly made an assumption that simplified our treatment of electric forces and fields. Coulomb's Law, and our subsequent treatment of electric potential, potential energy, and fields, only work in the nice simple manner described for stationary charges. (Gauss' Law, in its general form, still holds true for moving charged objects, but showing that is beyond the scope of this book.) Coulomb's Law assumed radial symmetry—the electric field lines spread out in a uniform manner in every direction, so that no matter what distance you are away from a charge, every point along that distance will experience the same electric field:

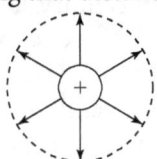

There's a problem, though, with the theory of special relativity, which was developed as a reflection on electromagnetism. (One of Einstein's famous papers published in 1905 was titled "On the Electrodynamics of Moving Charges.") Special relativity insists that information can't travel faster than the speed of light, which means that particles "causing" fields in different parts of space can't change those fields instantaneously. An electron placed on Earth is going to "create" an electric field that can be measured (however feeble it may be) on Pluto. (We're still using the words "cause" and "create" in quotation marks because causality and creation are not physical concepts. What we mean is that one measurable property of the electron is the electric field everywhere in space that would be present if that electron were the only charged particle in the universe.)

Suppose our electron starts moving. Information can only travel at the speed of light, and the distance from Earth to Pluto is great enough that even light would take a few minutes getting there—and therefore it would be a few minutes before the electric field on Pluto could possibly "know" that the electron had started moving. By the time this happens, the electron could have possibly already moved a great distance, depending on how fast it is moving. The radial symmetry of the electric field, however, will still be centered around the point at which the electron was placed originally, meaning that the electric field will *not* be radially centered around the electron itself, as Coulomb's Law specifies that it does. As the electron starts moving, the electric field will begin becoming displaced with it, and the rate at which the disturbance in the electric field propagates will be the speed of light. The electric fields are not radially centered around a moving charged particle, however, and this implies that Coulomb's Law is only a special case true for stationary particles. The more general expressions for electric (and magnetic) fields, called Jefimenko's Equations, were only worked out in 1966 by Ukrainian-American physicist, Oleg Jefimenko (1922–2009) and are a little too complicated to be worth using for computation here. Needless to say, the smaller the velocity of the charged particle, the better an approximation Coulomb's Law gives. Since the velocity of the particle is continuous (there is no sharp break between a velocity of 0 and a very small velocity), the perturbation of its electric field must be continuous as well.

Although we'd like to avoid many of the complexities that we would be presented with by trying to use Jefimenko's Equations and solve for the fields and potentials of moving charged particles, we can't remain bound to discussing stationary charges forever. Static electricity—the sort of thing that makes your hair stand up when you pass a comb over your head—is only a minor and relatively useless aspect of the electrical phenomenon, as far as any practical application goes. All electrical devices work through *currents* and *circuits*, and these involve charges moving. When things get too complicated, physicists step back and try to find a simpler way of thinking about the problem. For electrodynamics, the study of *moving* electric charges, our simpler approach to the problem will be taken by finding a new fundamental concept to start from, instead of electric fields from individual charges.

That basic starting point we want to begin with will be a composite concept called "current." Current is *not* the speed of a moving charge, despite common student misconceptions otherwise. The speed of a moving charge would not really be a new concept at all—all velocity, regardless as to how it is calculated, is still velocity, with units of m/s, and while new methods besides the kinematics equations may be found to calculate that velocity, the velocity itself does not form a conceptual starting point for deriving the physics of moving charges.

Current, instead, is defined as the *rate at which charge passes by a given point*, defined analogously to the manner in which current in fluid dynamics (the familiar "current" of a river or stream) is defined as the rate at which fluid passes by a given point. Current has been discussed since the first years of the 19th century, long before atomic and subatomic physics was understood, and consequently it's not necessarily the velocity of any single charge carrier, or individual particle carrying the charge. The charge could be carried by different types of particles with different charges on them, and different velocities, and nobody knew in the 1810s how big the particles carrying electric charge actually were. André-Marie Ampère, after whom the unit of current is named, had guessed that charge was carried by "electromagnetic particles," but he had no way of knowing how big those particles were, or how much charge they carried. Today we know that these particles are called electrons, and that they carry a charge $q = 1.6022 \times 10^{-19}$ C, an incredibly small amount of charge when compared with values seen on a day-to-day basis.

Current also has different units than velocity does. Two particles with the same charge and the same velocity passing by a given point will give us twice the current as one particle with that charge and velocity passing by the point, so the unit we need is not meters per second, but rather *coulombs* per second—the total number of coulombs that passes by a point in one second.

The unit of coulombs per second, in order to emphasize its fundamental role in electrodynamics, is given its own name—Ampère, or "Amp" (abbreviation: A) for short, after one of the early pioneers in electromagnetism, André-Marie Ampère. Current is nothing other than the charge divided by the time it took for that amount of charge to pass, so writing the definition in mathematical terms,

$$I \equiv \frac{Q}{t} \tag{11-1}$$

where Q is the total charge that passes by a given point in a time *t*.

Q is, by definition, the total *positive* charge passing by a given point. After all, unless we can distinguish discrete charged particles, the flow of positive charge in one direction is physically indistinguishable from the flow of negative charge in the other direction, going backward in time. (This indistinguishability is something we call a "symmetry," and this in particular is CPT symmetry, standing for charge–parity–time, where "parity" refers to the direction the charge is going. If all three quantities are reversed, then you end up with something physically identical to where you started. It wasn't until the 1950s that particle physics discovered examples of CPT symmetry being broken.)

In many advanced textbooks, current is only introduced *after* talking about magnetic fields and forces. There is good reason for this. Although an electric force can cause the charges to move through some electric potential, causing the current, a nonzero magnetic field centered around the current-carrying wire is something that can always be measured at some point in space as a property of the wire, just as a nonzero electric field is something that can always be measured at some point in space as a property of a charged particle. Currents and magnetic fields simply always go together. However, so long as we remember their inseparability, one new concept per chapter will be enough to handle. For pedagogical purposes, it is best not to calculate magnetic fields and forces at all for the present moment, and save it for the next chapter. For now, all that we need to remember is that one *does not have* a current without a magnetic field that is "created" in its vicinity by virtue of the charges moving.

In the next chapter, we see how to calculate the motion of charged particles under the influence of magnetic forces. Many circuits do in fact have magnetic forces causing the motion of charges—any wall socket coming from a power plant using an electromagnetic generator, for example. Most other circuits, including circuits powered by batteries, have the charged particles accelerating through electric potentials caused by electric charges, with a disparity of charge caused by something such as the difference in electronegativity between two substances. Circuits are either "AC" or "alternating current" circuits, where the direction of the charges oscillates rapidly, or "DC" or "direct current" circuits, where the direction of the charges is constant. Most wall outlets provided by electromagnetic generators employ alternating currents; these are conceptually more intricate and mathematically less elegant for computational purposes, and are not discussed in depth in this book. DC circuits are provided by batteries, and because they only involve simple linear motion like the kinematics and mechanics discussed in Chapters 3–7, calculations of the current or of the velocity of the charges will be just as simple and elegant as the kinematics and mechanics of which it is an extension. The force causing these charges to move is an electric force from a battery, and all of the circuits in the homework in this chapter are battery-powered DC circuits, so there is no need to understand the magnetic force in order to understand how to find the current.

It was mentioned above that the current is not the same physical quantity as the speed of the charge carriers. However, the definition of current relates the concept of charge (dealt with extensively in Chapter 9) with time, and all of the physics dealt with in Chapters 3–8 involved things moving—that is to say, changing position with time. Rewriting time as a function of velocity, we'll have a function for velocity in terms of current—allowing us to calculate the speed of the charge carriers whose current we have already measured.

Velocity, as always, is displacement over time, where the displacement of the electrons passing through a wire has a magnitude the total length of the wire traversed. Because displacement is still a vector, this length is going to be a *vector* **L**—a quantity identical to the displacement **Δx** used before, but emphasizing that the magnitude of the displacement through which the charges cross is the length of the wire they pass through. As it turns out, the velocity usually turns out to be remarkably slow, and for this reason the net forward velocity of charge carriers is usually called "drift velocity" or v_d. Since $v_d = \dfrac{L}{t}$, solving for t and substituting into the definition of current:

$$t = \frac{L}{v_d}$$

$$I = \frac{Qv_d}{L}$$

We're doing this because we want to find the drift velocity, v_d, so solve the equation for v_d:

$$v_d = \frac{IL}{Q} \qquad (11\text{-}2)$$

Q, the total charge passing by any given point in time t is unknown. However, since charge is carried by electrons, each of which has a charge with a magnitude $q = -1.6022 * 10^{-19}$, we can write the magnitude of the total charge as being the number of charge carriers times the charge per carrier, or $Q = qN$. We say "magnitude" because electrons actually have a negative charge, and therefore move from a lower to a higher potential—and uppercase Q is, by definition here, the total *positive* charge flowing past a given point. (Capital letters are usually used to the "total" quantity, while lowercase letters for the individual quantities.) We can imagine fictitious "charge carriers" being the positively charges or "holes" left by valence electrons in the atoms they leave. These "holes" travel from atom to atom—the overall positive charge left by the empty spot will attract an electron from a neighboring atom, filling that hole but leaving another one next door.

Calculating the charge would be much less confusing if we had defined current in the direction of motion of the actual charge carriers, the electrons. However, "positive" charge was defined as such by Benjamin Franklin, who did not know which way the charges were moving, and simply took a guess—and guessed wrong. A positive current is defined as the direction of positive charges; a negative current actually tells you the way the electrons are moving.

In order to calculate Q in equation 11-2, we need to find the total number of charge carriers, N. We can assume that the total number of electrons (or "holes") is usually going to be equal to the number of atoms. Normally, only one valence electron is transferred from one atom to another. The number of atoms in the segment of wire with length L through which the electrons pass in time t is the *number density* of the atoms times the volume. *Number density* is the number of things per volume; just as mass density ρ is mass per volume, number density is number per volume. Number density is traditionally denoted by a lowercase n, while the total number of items is traditionally denoted by a capital N. The number density of charge carriers for monovalent charge transfer is going to be the same as the number of atoms, an assumption we can usually make.

Wires tend to be cylindrical, and a cylinder of radius r and length L has volume $V = \pi r^2 L$, so

$$N = n\pi r^2 L \qquad (11\text{-}3)$$

Since $Q = qN$, we can rewrite the Q in the equation for drift velocity as

$$v_d = \frac{IL}{qn\pi r^2 L}$$

The Ls cancel—the speed of the charge carriers, as one would expect, not dependent on the length of the wire through which they are passing, and the formal, "useful" equation to calculate drift velocity becomes

$$v_d = \frac{I}{qn\pi r^2} \qquad (11\text{-}4)$$

The tricky part here is calculating n. The charge q is always $1.6022 * 10^{-19}$. Number density isn't something most handbooks or encyclopedias will simply tell you, however, instead, they'll tell you the *mass* density ρ, found in Table 11-1 on page 301. However, one can find the number density if given the mass density, using simple algebra.

Number density is the number of atoms per volume, and

$$\frac{\text{Number of atoms}}{\text{volume}} = \frac{\text{mass}}{\text{volume}} \times \frac{\text{number of atoms}}{\text{mass}}$$

The first quantity on the right side of the equals sign, mass per volume, is mass density ρ. The second term is the number of atoms in a kilogram. This isn't something necessarily easy to find directly either, but the number of atoms in a kilogram is 1 divided by the mass of the atom in kilograms:

$$\frac{\text{number of atoms}}{\text{mass}} = \frac{1}{\frac{\text{mass}}{\text{atom}}}$$

The mass of each atom can be found from a periodic table. The atomic weight of each element gives the mass of a *mole* of atoms in *grams*. A mole of atoms is 6.022 * 10²³ atoms (that's Avogadro's Number, given the label N_A), so dividing the atomic weight by N_A gives the mass of a single atom. The mass given on the periodic table needs to be converted from grams to kilograms. Chemists, for whom the periodic table is most useful, use the "cgs" or "centimeters–grams–seconds" system of standard units, whereas physicists use "mks" or "meters–kilograms–seconds." One kilogram is 1,000 grams.

If the atomic weight *in kilograms* is given the symbol w (a convention chosen for this book), then the weight of a single atom is $\frac{w}{N_A}$, and therefore the number of atoms in a kilogram is $\frac{N_A}{w}$. It's this quantity that needs to be multiplied by the mass density in order to get the number density:

$$n = \rho \frac{N_A}{w} \tag{11-5}$$

For example, suppose we want to find the drift velocity of electrons through an iron wire. Iron has a mass density $\rho = 7874$ kg/m³, according to Table 11-1, and it has an atomic weight 55.85 g/mol = 0.05585 kg/mol. Avogadro's Number is 6.022 * 10²³ atoms/mol, so substituting these numbers into equation 11-5, we find that the number density of iron is (7874)*(6.022*10²³)/(0.05585) = 8.49 * 10²⁸ atoms/m³.

Resistivity, Resistance, and Ohm's Law

In order to calculate the drift velocity, one has to not only know the number density of charge carriers, but also the current. The current at any point is measurable easily enough, but physics is never comfortable with simply saying "you can measure it"—the assumption of a rational, predictable universe demands that we be able to *calculate* the value of the current, if the current is indeed predictable and not simply arbitrary. Fortunately, by thinking through the situation clearly and describing an assumption called "Ohmic behavior" as a mathematical formula, we can find a simple relationship between the current between two points and the electric potential difference across which the charges are passing.

As you recall from Chapter 10, charged particles placed in an electric field will experience an electric force, causing them to move. The electric force on a charge q is $\mathbf{F} = q\mathbf{E}$, and if the force causes the particle to move across a distance $\mathbf{r}$, it will have traversed a potential difference $V = \mathbf{E} * \mathbf{r}$. The potential difference V is what we deal with most directly. The potential difference induced between the terminals of a battery is printed on the battery. As he writes this, the author is holding in his hands a 1.5V size AAA battery, a 1.5V size AA battery, a 1.5V size D battery, and a 9V pile alkaline, all common household items. A wall outlet will typically provide a much greater potential difference, one standardized by country. Most wall outlets in the United States provide a potential difference of 120 V. We know that potential differences cause charges to move—and the rate at which charge passes by a given point is the current. We'd like to know how to calculate the current when given the potential difference.

It would seem a reasonable assumption that for any given wire, the current is going to be proportional to the electric field induced at a given point. Between two points separated by a distance $\mathbf{r}$ with a potential difference V induced, there will be a uniform electric field of magnitude $E = \frac{V}{r}$ induced, as we already know. The current is going to be proportional to E

rather than V, since a high potential difference induced across a short distance is going to pull a much stronger current than that same potential difference across a much longer distance. One can picture V as the height of a hill with r being the horizontal displacement traveled from the top to the bottom—the steeper the hill, the faster one will accelerate down it. And the reason why this analogy works is because the potential difference is proportional to the electric charge on each end creating that potential difference—and the closer two charges are, the stronger the electric force between them, as we know from Coulomb's Law.

I is only proportional to E when one is comparing different currents and electric fields in the same wire, however. The same electric field will induce a stronger current in a thicker wire than in a thinner one, since more electrons will experience the same electric field. We can assume that the current is going to be uniformly distributed in the wire. There needn't be any irregularity or disuniformity in a wire causing charge to move faster on one part than another, or causing the charge to move faster in one direction than the other. Of course, there *are* anisotropic materials for which this assumption isn't going to hold true, and in which the electric field and proportionality constant will become tensors, a more conceptually challenging mathematical object than is appropriate for discussion in this course. So, ignoring difficult situations like that, we can generalize our statement earlier (that *for any given wire* the current will be proportional to E) by saying that for *all wires* the *current density* is proportional to E. The *current density* is current divided by area, and since we used the symbol I for current, the next letter in the alphabet is J, and current density is consequently defined as

$$J \equiv \frac{I}{A} \qquad (11\text{-}6)$$

And we will maintain the assumption that J is proportional to E.

This assumption is not actually always correct. When a strong enough electric field is applied, the number of electrons being pulled reaches a maximum. There are only a finite number of electrons available, and all of these electrons repel each other—and they will in fact collide frequently as they're being pulled down a wire, especially when they start getting crowded. So this assumption only works below a certain maximum E. For electric fields at which it works, the material in question is said to obey "Ohmic" behavior after the late 18th- and early 19th-century physicist Georg Ohm. "Non-Ohmic" behavior is an interesting field of study for electrical engineers, but is not something we need to concern ourselves with in this course.

These collisions between electrons, for Ohmic or non-Ohmic behavior, are the only reason why the current is constant. If a force is being applied to the charges, then they should be accelerating, and the current should therefore be increasing with time. However, the acceleration each charge experiences from the electric field is counteracted by an equal and opposite deceleration when it hits a particle in front of it. David Griffiths[1] uses the helpful analogy of cars driving down a street with many stoplights. No matter how fast we accelerate in between the stoplights (collisions), our average velocity is going to be the same, and despite the attempts of many drivers on the road, one is not actually going to get to their destination any faster. A more detailed explanation, a bit too technical for and beyond the scope of this book, can be found in Griffiths's now-classic textbook.[2]

Unfortunately, there is no way to simply calculate the proportionality constant between J and E. If they are linearly proportional as we are assuming (defining any behavior for which this assumption is true as "Ohmic behavior"), then they can be written as a simple linear equation: J equals some constant times E. However, valence electrons in different materials are bound with different strengths, depending on that material's electronegativity. We can't write a

---

[1] David Griffiths, *Introduction to Electrodynamics*, 3rd ed. (Upper Saddle River, NJ: Prentice-Hall, 1999), p. 289.
[2] Ibid.

nice and simple equation that applies universally, and even early attempts to derive conductivity using statistical thermodynamics have been superceded by more modern models employing quantum mechanics. All of these models involve physics, which has not been introduced in this book. Instead, we will simply do what Georg Ohm did, and make up a name for the proportionality constant and measure it in the laboratory. The proportionality constant is called *electrical conductivity* and is denoted by the Greek letter sigma:

$$J = \sigma E \qquad (11\text{-}7)$$

This equation is called "Ohm's Law". E is only the *magnitude* of the electric field, because J (like I) is a scalar. Ohm's Law is simply an empirically determined proportionality assumption, one that holds true under some circumstances (called "Ohmic behavior") but not others (called "non-Ohmic behavior"). It is not a universally true principle that holds true for all materials under all circumstances—just the majority of circumstances seen in everyday life, including the homework problems we'll see in this book.

This isn't the most well-known form of Ohm's Law, however, and not the form usually used for calculations. V is usually what is given to us (from the label on the battery, or from the wall outlet), rather than E, and we want to know the actual current I rather than J. But now that we have Ohm's Law, we can rewrite E in terms of V and J in terms of I to get something more useful. As we know, $E = \frac{v}{r}$, and since r is the length of the wire, we might as well rewrite it as $E = \frac{v}{L}$. Rewriting Ohm's Law thus,

$$\frac{I}{A} = \sigma \frac{V}{L}$$

and solving for V,

$$V = \frac{L}{\sigma A} I \qquad (11\text{-}8)$$

The proportionality constant $\frac{L}{\sigma A}$ is called "resistance" and is given the symbol R. Unlike conductivity, which is a property of the *material* regardless of size or shape, and is therefore something that can be found from a table, *resistance* is a property of the actual physical object. A longer or bigger object is going to have a greater resistance than a smaller object made out of the same stuff. We solved for V rather than I because Ohm's Law is traditionally written that way:

$$V = IR \qquad (11\text{-}9)$$

Resistance and current are inversely proportional. Given some potential difference V, the greater the current, the less the resistance—and vice versa. Resistance is called such because it "resists" the current, impeding it (in more complicated circuits, with alternating currents and things like inductors and capacitors, it will actually be called "impedance"). In honor of Georg Ohm, the unit of resistance (Volts per Ampere) is called an "Ohm," symbolized by the Greek letter Omega (Ω).

Resistance is proportional to the length of a wire because electrons have a greater distance over which they will be repelled by other electrons in the wire and see their current "resisted." It is inversely proportional to the cross-sectional area of a wire because the average velocity (or "thermal velocity") of the electrons is in the direction of the current—and collisions causing resistance will occur by electrons hitting each other in that direction. The more electrons one has moving side by side, the greater amount of charge will be passing by that point, and the repulsion between electrons side by side with each other does not contribute to resistance, so the current is increased. Finally, the resistance is inversely proportional to conductivity.

Instead of talking about conductivity, which is a measure of how well a material conducts electricity, we might want to talk about "resistivity," or how well a material resists electricity.

Resistivity is also a property of the *material* rather than of the object, and is traditionally denoted with the symbol ρ (as is mass density, although they are unrelated concepts), and is simply the reciprocal of conductivity:

$$\rho \equiv \frac{1}{\sigma} \qquad (11\text{-}10)$$

Defining resistivity this way, we can write resistance as a function of resistivity. The unit of resistivity is an Ohm-meter (Ω m).

$$R = \rho \frac{L}{A} \qquad (11\text{-}11)$$

Resistance is properly speaking, of course, a property of a *system* rather than of an "object." There is no single, set meaning for an "object" or "thing" in physics except for a point particle, and point particles certainly do not have length or cross-sectional area, or resistivity for that matter, and therefore to speak of "resistance" for point particles is meaningless. A point particle is an extended object, a section of wire through which the charges pass, and since the resistance of each section of wire is proportional to its length, and the total length of a wire is the sum of the lengths of its parts, the total resistance of a wire is going to be the sum of the resistances of its parts. The total resistance of a string of wires connected to each other is called the "equivalent resistance," or R_{eq}, because we are treating the contributions from each of the parts of the wire—each of which is a resistor in its own right—as contributing to the resistance of one, single resistor. Any time an electron passing through a circuit must leave one resistor (one section of wire) and enter another one, those two resistors are said to be in "series." Graphically, a resistor is represented by a zig-zag, with the different resistors connected by straight lines representing wires (although, in reality, the resistance will be distributed through the entire wire, rather than localized in one place):

—⋀⋀⋀—⋀⋀⋀—⋀⋀⋀—

The three resistors shown above are in series; electrons passing through one must pass through the next one sequentially, without having any other option. These could be three light bulbs strung in a row, or these could be three parts of the same wire, or any configuration in which the current through each resistor is the same. Because electric potential is a function of displacement and not path length, and because the electrons are not doubling back on themselves as they pass through the three resistors, the total potential difference traversed by the electrons is also going to be the sum of the potential differences traversed through each resistor individually, and we can write

$$V_{tot} = V_1 + V_2 + V_3 + \ldots V_n$$

for up to n resistors (only three are shown here). Ohm's Law tells us that $V = IR$, and the current I is going to be the same for each resistor since there is nowhere else for charges to go, and so writing the total potential difference as being that current times some "equivalent resistance" (the total resistance of all the individual objects treated as a single whole),

$$IR_{eq} = IR_1 + IR_2 + IR_3 + \ldots IR_n$$

Since the current is the same in each term, we can divide each term by I, leaving us with

$$R_{eq} = R_1 + R_2 + R_3 + \ldots R_n \qquad (11\text{-}12)$$

for resistors in series, just as argued above from the fact that the total length of the wire will be the sum of the cumulative lengths of each part.

Sometimes, electrons will not be forced to travel through different resistors sequentially, but rather face a branch in the circuit—called a junction—and be forced to take one branch or the other, much like water in a river with branches being forced to travel down one branch or another, but not both. All circuits are going to eventually return to the same place that they started, since the lower and higher potential plates in a circuit are both going to be on the same battery, and the branches are going to have to recombine. The potential they return to after recombining must be the same no matter which branch is taken—since potential difference, by definition, is the work done by a *conservative* or path-independent force—and therefore the potential differences across each branch must be the same.

Because the current splits and the different branches are usually drawn side by side, whenever resistors have the same potential across them, they are said to be in "parallel." Just as we did with resistors in series, whenever a branch only contains resistors, we can find an "equivalent resistance" for the resistors in parallel.

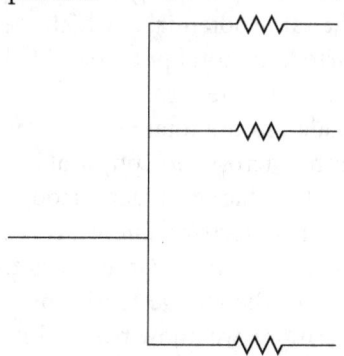

The total current going through all the branches together must equal the sum of all the individual currents, since conservation of energy and conservation of mass together imply conservation of current (current carrying charges aren't going to appear or disappear out of thin air, and without a force acting on them, the velocity of the charges isn't going to increase or decrease). If you take the *total* current going through all the resistors together, and the potential difference across any individual branch, and solve Ohm's Law for the resistance using that current and potential difference, the resistance found would be the "equivalent resistance" for resistors in parallel. We can find a direct expression for this equivalent resistance by solving Ohm's Law for current and writing the total current as the sum of all the individual currents, giving us

$$\frac{V}{R_{eq}} = \frac{V}{R_1} + \frac{V}{R_2} + \frac{V}{R_3} + \ldots \frac{V}{R_n}$$

Since V is the same for each branch, or for all of them, it appears in every term and can be canceled out, leaving us with

$$\frac{1}{R_{eq}} = \frac{1}{R_1} + \frac{1}{R_2} + \frac{1}{R_3} + \ldots \frac{1}{R_n} \qquad (11\text{-}13)$$

for the equivalent resistance of n resistors in parallel.

Kirchhoff's Rules

The reason we want to know Ohm's Law is usually so we can calculate the current running through any given resistor. However, when resistors are part of complicated circuits including multiple objects, in order to know the current going through each resistor using Ohm's Law we need to know the potential difference the charges cross from one end of the resistor to the other, and the potential difference across a resistor is not something that is going to be calculated quite as easily.

The means for figuring out the potential difference across resistors was developed by 19th-century German physicist Gustav Kirchhoff. The method is so simple and commonsensical that the equations are often referred to as "Kirchhoff's Rules," rather than "Kirchoff's Laws." There are two such rules—the "loop rule" and the "junction rule"—which, like Newton's Laws, are written out as equations used to solve for the currents, which are left as unknown variables.

Kirchhoff's "Loop Rule" is based on a simple fact about potential. Electric potential is a scalar field, meaning that one can measure an electric potential at every point in space. Given any set physical situation and distribution of electric charge causing a nonuniform potential, the potential is only a function of coordinate location—*any* object coming to any given point, no matter what path it took, is going to come to the same electric potential. Any charged object leaving one point and taking any path that returns to that point is going to return to the same electric potential.

This fact is useful for closed direct-current circuits, like many battery-operated devices, as well as any devices that operate on solar-generated electrical power. If charge carriers ("holes," or fictitious positive charges moving in the direction of a current) leaving one point in the battery out of the positive terminal end up coming back to that same point through the negative terminal, they will return to the same potential at which they started. If a charge returns to the same potential at which it started, the total potential difference across the entire path that it takes—across the entire circuit—must be zero.

Potential difference is a scalar, so the total potential difference across the entire circuit is the sum of the potential difference across each physical object that the charge carrier encounters. Most physical objects that the charge carrier encounters are devices such as light bulbs, which employ electrical energy. For electrical energy, $U = qV$, the electrical energy per charge carrier consumed by these devices is that charge times the potential difference across that device (the potential difference of the charge carrier between the points before and after it encounters the device). Furthermore, we know from Ohm's Law that the potential difference across any device with a current running through it is given by $V = IR$. Electrons themselves are not consumed—that would violate conservation of mass, the idea that matter is neither created nor destroyed; rather, what is consumed is the electrical *energy* of the electrons being used. Since the electrons are not being consumed, there must be an equal number of electrons leaving as entering the electrically powered device, which requires that a current be passing through the device—and therefore, the device has some resistance.

Devices with resistance are called "resistors," and since the resistance of an object depends only on its composition (which determines ρ) and its dimensions, the resistance can be determined very easily experimentally by connecting it to a simple circuit that only includes a battery and either an ammeter (device that measures current) or a voltmeter (device that measures potential difference between two electrical leads), and using Ohm's Law, or by knowing its composition and measuring its dimensions and using equation 11-11. Because it is very easy to determine the resistance experimentally, this will be usually treated as a "given" in the homework problems, leaving the current as a variable for the student to solve for.

Kirchhoff's Loop Rule says that the sum of potential differences across any closed path or loop must be 0. Written mathematically, this is expressed as

$$\Sigma V_{loop} = 0 \qquad (11\text{-}14)$$

To use this rule in order to find the current running through a resistor, start by drawing *any* closed path or loop that includes the resistor in question. The loop must end at the same point that it begins, and in order to avoid confusion with signs (knowing whether to add or subtract potential differences), it is best not to backtrack and go backward through a segment of wire one has already gone forward through. Then one should write an equation, starting with a battery and tracing one's path through that loop. Each time an object is encountered, the potential difference across that object should be written down and added to the previous potential differences. The potential difference across each battery is given as the voltage across that battery; the potential difference across each resistor is $-IR$. The potential difference across each resistor is negative because a resistor, by *using* electrical energy, brings a charge carrier down to a *lower* electric potential, until the electric potential is down to 0, at which point all the electric potential energy has been used up. For our problems, assume that, no matter how long the wires connecting the resistors are, they will be resistanceless. (In real life, we can't make that assumption—and experimentalists running on a low budget will use long packets of wires to run currents through in order to lower the potential of a circuit—but the resistance of a wire can be easily calculated once we know the resistivity of its material and its length and cross-sectional area.)

For example, the next image shows a simple circuit with two batteries (indicated by the longer line representing the positive terminal of the battery and the shorter line indicating the negative one) and a resistor (denoted by the zig-zag).

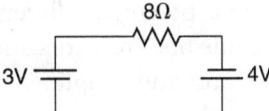

By convention, when drawing the loop we usually start at the longer line of the battery, include the potential difference across that battery, and proceed in whichever direction is "out" of the battery until we get back where we started. If we encounter more batteries, we *add* the potential difference from each battery when we come out of the longer line, and we *subtract* the potential difference from each battery when we come out of the shorter line. The potential difference across the resistor is IR, so Kirchhoff's Loop Rule for this circuit will be

$$3 - 8I + 4 = 0$$

This can be easily and quickly solved for *I*, giving us *I* = 7/8 Amperes.

If the 4V battery had been turned around, like this, we would have to subtract the 4V battery because it is turned around in the direction against the other battery:

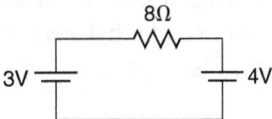

The loop rule in this case gives us

$$3 - 8I - 4 = 0$$

$$I = -\frac{1}{8} \text{ Amperes}$$

The second rule is called "Kirchhoff's Junction Rule." A *junction* is any point in the circuit where a charge carrier has two or more options as to which way it will end up going; it is analogous to a fork in a river where the river splits. And like a fork in a river where the total amount of water coming into a fork equals the sum of the amount of water leaving through each branch of the fork, conservation of mass—and its derivative, conservation of charge—requires that the total amount of current coming into a junction must equal the total amount of current leaving a junction. In simple mathematical language,

$$I_{in} = I_{out} \tag{11-15}$$

A junction is the only thing in a circuit that will change the current. A resistor will decrease the *potential* of the charge carriers passing through it, but *not* the current—a junction, by contrast, does *not* have any potential difference (if the wires are resistanceless), but the current on each branch connecting to the junction is going to be different. A junction is not a stretch of wire or an object, just a single point where the wires are connected.

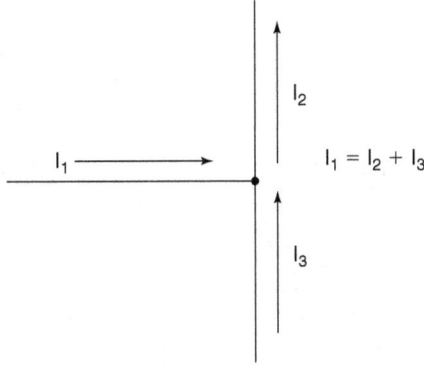

To solve for the currents in complicated circuits, one should start by identifying the resistors in the circuit. If two resistors are in series, there will be no junction between them (because to be in series implies having the same current), so one can find the equivalent resistance between them. One can also find the equivalent resistance of two resistors in parallel; however, it is the entire *branch* that has the same potential difference as the branches that it is in parallel with—and these branches may include both resistors and batteries in series together.

After one has identified the resistors and simplified it (if necessary) by finding the equivalent resistances of items in series, one should give names to the currents running through these resistors, beginning with I_1, I_2, and so on up to I_i for *i* resistors. One cannot algebraically solve for more variables than one has equations, so in order to find the currents, as many equations as currents are required. Unfortunately, these equations will often be *coupled*—meaning they will contain two or more variables each, requiring them to be solved simultaneously using substitution or elimination.

To begin writing the equations, start from a battery (any battery) and begin drawing loops. Draw loops until every object on the circuit has been covered once, writing down the loop rule for each loop. Usually this will provide fewer equations than one needs, and drawing more loops composed of smaller ones superimposed on each other will not help (solving it will end up giving 0 = 0). To provide the equations that are missing, pick as many junctions as you need equations, and solve the junction rule.

For example, consider the following circuit. There are four resistors, and none of them are in series, so there are four different currents that need to be solved for. Three loops cover the entire circuit, and one more equation is needed, so three loop rules and one junction rule are needed to solve for the currents:

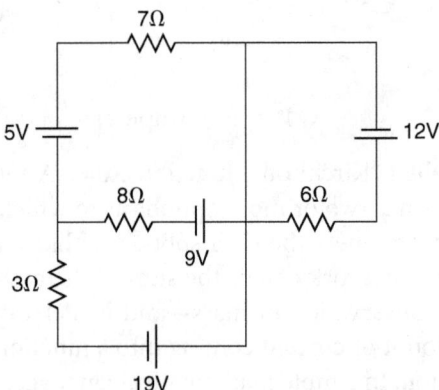

The first loop can start from the 5 V battery and proceed out of the positive terminal (the longer line), going in a clockwise fashion through the 7 Ω resistor, into the positive terminal of the 9 V battery (which will need to be *subtracted*), through the 8 Ω resistor, and then back up to the 5 V battery again. We can call the current going through the 7 Ω resistor "I_1" and the current going through the 8 Ω resistor "I_2," specifying that both proceed in the direction in which we have defined the loops:

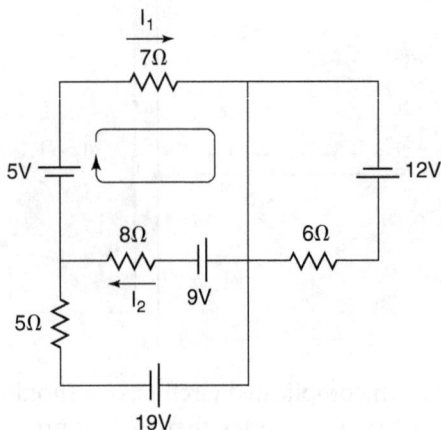

Because Ohm's Law tells us that the potential difference across a resistor is IR and the potential difference across the batteries is already given, the loop rule for this loop will be written

$$5 - 3I_1 - 9 - 8I_2 = 0 \tag{11-16}$$

This has two variables, so a second loop containing either I_1 or I_2 needs to be written out.

For a second loop, we can start from the 19 V battery on the bottom, and proceed counterclockwise into the positive terminal of the 9 V battery, through the 8 Ω resistor, and then down through the 5 Ω resistor. We have already labeled the current through the 8 Ω resistor I_2, which we specified was going in the direction that we are still going now (to the left); a name I_3 needs to be given to the 5 Ω resistor. Following the direction of the loop, I_3 can be oriented downward. We will therefore be tracing a loop going in this direction:

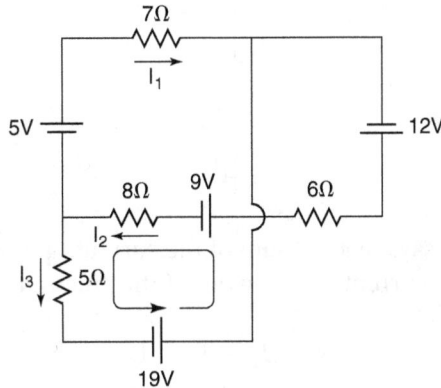

Writing out the loop rule starting at the 19 V battery,

$$19 - 9 - 8I_2 - 5I_3 = 0 \tag{11-17}$$

We now have two equations written down, including three unknowns. The fourth loop is simple—it only has one battery and one resistor, and therefore only one current going through the resistor to solve for—but the current is going to be I_4, something different than the other three. We can start from the 12V battery and go clockwise around the loop:

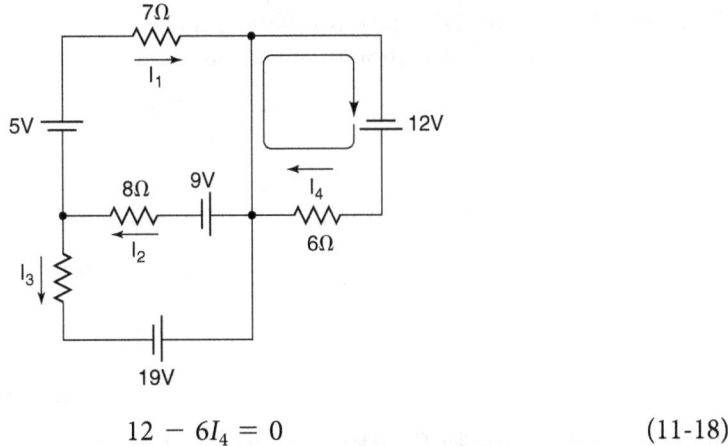

$$12 - 6I_4 = 0 \tag{11-18}$$

Equation 11-18 can be solved immediately for I_4, giving us $I_4 = 2.0$ A.

Equations 11-16 and 11-17 have three variables between them, and one more equation containing some combination of those three variables is needed to finish solving the problem. We've already covered the entire circuit with loops, and no new circuit is going to give any new information. We must finish the problem by employing the junction rule. There are three junctions, indicated by large dots in Figure 11-10, and while any of these can be used correctly to solve the problem, we don't know the name of the current between the two junctions on the right side. The current changes at each junction, and since there is no resistor between the two junctions on top and on bottom on the right side of the figure, we don't have a name for it—and introducing another name for a current introduces another variable that needs solving for, and *both* junctions will end up having to be used to find the requisite number of equations. That's harder than it needs to be. I_1, I_2, and I_3 all proceed into or out of the junction on the left, making that junction the easiest one to use.

Note that in the figures, arrows were drawn indicating the direction of the currents. I_1 proceeds to the right through the 7 Ω resistor, meaning that if we trace its direction back to the junction where it comes from, it comes *out* of the junction on the left. I_2 crosses the 8 Ω resistor going to the left, going *into* the junction, and I_3 comes down through the 5 Ω resistor, coming *out* of the junction:

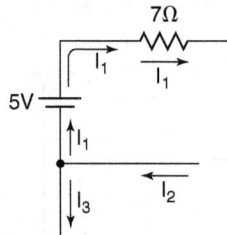

Kirchhoff's Junction Rule says that the sum of the currents going into a junction (just I_2, here) must equal the sum of the currents coming out of the junction ($I_1 + I_3$), so the junction rule is written out as

$$I_2 = I_1 + I_3 \tag{11-19}$$

Solving for I_1, I_2, and I_3 is just algebra. Equation 11-19 is already solved for I_2, so we can eliminate I_2 in equations 11-16 and 11-17 by substituting $I_1 + I_3$ for I_2:

$$5 - 3I_1 - 9 - 8(I_1 + I_3) = 0$$

$$-4 - 11I_1 - 8I_3 = 0 \tag{11-20}$$

$$19 - 9 - 8(I_1 + I_3) - 5I_3 = 0$$

$$10 - 8I_1 - 13I_3 = 0 \tag{11-21}$$

Equations 11-20 and 11-21 have two variables and two unknowns, and can be solved using substitution, with either equation being solved for either variable first:

$$8I_3 = 4 + 11I_1$$

$$I_3 = 0.5 + 1.375I_1 \tag{11-22}$$

$$10 - 8I_1 - 13(0.5 + 1.375I_1) = 0$$

$$10 - 8I_1 - 6.5 - 17.875I_1 = 0$$

$$3.5 - 25.875I_1 = 0$$

$$I_1 = 0.135 \text{ A}$$

Substituting this value for I_1 back into equation 11-22,

$$I_3 = 0.5 + 1.375(0.135) = 0.686 \text{ A}$$

Substituting both of these values into equation 11-19 to find I_2,

$$I_2 = 0.821 \text{ A}$$

Since all of these values are positive, the current (flow of *positive* charges) is flowing in the direction we indicated on the diagrams in each case.

Capacitance

In the last chapter subheading, we explored circuit analysis with just two types of objects: batteries (which provide power) and resistors (which consume it). We'll talk more about what exactly electrical power is and how to use it in the last subheading of this chapter. Now we shall add a third type of object, a capacitor. Capacitors were discussed at some length in Chapter 10, where capacitance was introduced and Gauss' Law was used to find the electric field inside a capacitor. Capacitors have a wide variety of applications in direct-current circuits because they take regular intervals to charge and discharge. A windshield wiper on a car, for example, discharges while it is wiping, and charges between wipes.

Recall that the definition of capacitance is $C = \dfrac{Q}{\Delta V}$ where Q is the total charge stored on the positive plate of the capacitor, while ΔV is the potential difference between the plates. Kirchhoff's Rules asks us to add together the potential differences between plates on a capacitor, so solving for ΔV we see that

$$\Delta V = \frac{Q}{C} \qquad (11\text{-}23)$$

The capacitance is an easily measurable quantity that will usually be given to us. Most capacitors have the capacitance printed on the object, but Q is a bit more tricky to find. Writing out Kirchhoff's Loop Rule for a simple circuit with a battery, a capacitor, and a resistor in series gives us

$$V - IR - \frac{Q}{C} = 0 \qquad (11\text{-}24)$$

While Q is a variable that needs to be solved for, this can't be done in a single equation because I will now be time-dependent.

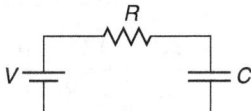

Such a circuit is called an "RC" circuit because of the presence of both resistors and capacitors.

A capacitor placed inside of a circuit is initially just a couple of metal plates capable of maintaining charges on the two plates without it shorting out (neutralizing itself by sparks jumping across the plates) up to some given voltage difference induced between them. The charges do not come on the capacitor automatically, packaged in the box one purchased it in. They have to be deposited on the plates by connecting the capacitor to a closed circuit, with the positive terminal of the battery drawing electrons away from the positive plate of the capacitor and the negative terminal depositing electrons on the negative plate.

When the capacitor is first connected, there are no charges on the plates, so the potential difference between them is 0, and the circuit can be analyzed as if it were not there. In the simple circuit in the diagram above, Ohm's Law tells us that the current passing through the resistor would be $I = \dfrac{V}{R}$ where R is the resistance of the resistor in series with the capacitor and V is the potential difference created by the battery and/or anything else on the loop. However, if the circuit is left to charge the capacitor up indefinitely, the potential difference between the plates will eventually become equal and opposite to the potential difference between the terminals of the battery. The potential difference across the capacitor then becomes $-V$, and Kirchhoff's Loop Rule for this circuit becomes

$$V - IR - V = 0$$

Since the resistance R is fixed, the current I must become 0.

A formula for I as a function of time cannot be found without resorting to calculus, and the derivation (presented here) can be omitted by the student unfamiliar with calculus. The derivation relies on solving $V - IR - \frac{q}{c} = 0$ for Q, relying on the fact that (in general) instantaneous current is $I = \frac{dq}{dt}$. (The definition given in the text here, $I = \frac{Q}{t}$, is average current; the definitions yield the same results when the current is steady. By convention, capital Q is usually used for a single value, or for the *final* value of the charge deposited on one of the plates, whereas lowercase q is generally used for a time-dependent variable; hence my otherwise inexplicable shift here to lowercase for the derivation that follows.) This is a separable differential equation that can be solved by isolating the $\frac{dq}{dt}$, separating the function putting everything with "q" on one side and everything with "t" on the other, and integrating:

$$V - \frac{dq}{qt}R - \frac{q}{c} = 0$$

$$V - \frac{q}{C} = \frac{dq}{dt}R$$

$$\frac{dq}{dt} = \frac{V}{R} - \frac{q}{RC}$$

In order to be able to integrate a differential equation, each side must have just *one term* with all q's and no t's, or all t's and no q's. In order to get the right side as one term, put both terms over a common denominator:

$$\frac{dq}{dt} = \frac{CV}{RC} - \frac{q}{RC}$$

$$\frac{dq}{dt} = \frac{VC - q}{RC}$$

Separate the equation by dividing both sides by $CV - q$, and by multiplying both sides by dt:

$$\frac{dq}{CV - q} = \frac{dt}{RC}$$

Using substitution, let $u = CV - q$ and therefore $du = -dq$. Integrating this in terms of u,

$$\int_{VC}^{VC-q} -\frac{1}{u} du = \frac{1}{RC} \int_0^t dt$$

with apologies for the lack of rigor in using lowercase t and q for both the differential and the limit of integration.

The integral on the left side gives a natural logarithm. Using a property of logarithms, $\ln(CV - q) - \ln(CV) = \ln\left(\frac{CV - q}{CV}\right)$; the expression on the left-hand side will be negative of this (because of the negative sign in the integrand). The integrand on the right side is just t, which is then multiplied by the constant out front, so the solution to this differential equation is

$$-\ln\left(\frac{CV - q}{CV}\right) = \frac{t}{RC}$$

Dividing each term in the argument of the natural logarithm by CV and moving the negative sign over to the other side,

$$\ln\left(\frac{1 - \frac{q}{CV}}{1}\right) = \frac{-t}{RC}$$

Raising the natural logarithmic base e to both sides of the equation,

$$1 - \frac{q}{CV} = e^{-t/RC}$$

Solving for Q,

$$1 - e^{-t/RC} = \frac{q}{CV}$$

$$q = CV(1 - e^{-t/RC}) \tag{11-25}$$

Here q is the final charge measured at some point in time, whereas CV is the final charge on the positive plate of the capacitor after a very long time, as t approaches infinity, which previously we have been calling capital Q.

The current, as a function of time, is found by redifferentiating this separated equation with respect to time. Current is by definition $\frac{dq}{dt}$, and taken the time derivative of the right-hand side of the equation will give us a time-dependent function for I. Here, CV is rewritten as Q:

$$I = \frac{dq}{dt} = \frac{Q}{RC}e^{-t/RC}$$

But $\frac{Q}{C}$ is just V, so

$$I = \frac{V}{R}e^{-t/RC} \tag{11-26}$$

It was stated earlier that if a capacitor was charged for an indefinitely long period of time, the potential difference across the capacitor would become V, with no current running through the resistor.

At that point in time, the capacitor is physically indistinguishable from the battery. If one were to remove the battery, there would be no "push" canceling out the "push" from the capacitor, and a current would again start to flow—backward—from the capacitor. The difference between the current coming from the capacitor and the current coming from the battery is that, with the only other object in the circuit being a resistor, the battery would provide a steady current whereas the capacitor's current would decay over time, until all of the electrons on the negative plate of the capacitor had fully discharged by recombining with the cations on the positive plate. The current's behavior over time is in fact physically identical, except backward, from its behavior when the capacitor was charging, and by writing I as a derivative again, recognizing that the capacitor provides a potential difference of *negative* $\frac{q}{c}$ because it is pushing charges the opposite direction than it had been before, and solving and then differentiating $-\frac{q}{c} - \frac{dq}{dt}R = 0$ gives us

$$I = -\frac{Q}{RC}e^{-t/RC} \tag{11-27}$$

which is identical, except for being opposite, to Equation 11-26. The remaining steps to the proof for this are trivial, and are left to the reader.

When multiple capacitors are present, Kirchhoff's Rules will usually have to be employed to solve for the charge on each plate, just as was done before. Equation 11-24 needs to be interpreted carefully, since the *V* in the equation refers to *everything* except the potential difference across the capacitor and across the resistor in series with the capacitor. If the loop contains more than just a battery, a resistor, and a capacitor, the potential difference from everything except the capacitor and the resistor in series with it need to factor into *V*. An example of this will be worked through by the student in problem S in this chapter.

There are times, however, when RC circuits can be simplified by finding the "equivalent capacitance" of capacitors in series or in parallel. Capacitors in series are, like resistors in series, hooked up so that there are no junctions between them. They differ from resistors in series; however, since charges can't actually cross the gap from one plate of a capacitor to another, a capacitor by definition is something that keeps charges separate from each other.

What happens in capacitors in series is that the current causes an overall positive charge to develop on the plate of the capacitor closest to the positive terminal of the battery. (One can think about them as positive charges or "holes" coming from the battery; of course, in reality, the process works backward, with negatively charged electrons being emitted from the negative terminal of the battery.) That positive charge on the capacitor attracts electrons from the positive plate of the next capacitor over, pulling them to the negative plate of the first capacitor, and the positive charge induced on the second capacitor induces a positive charge on the third one, and so on:

What is important to notice is that because the charges on each plate are all being induced by the other charges, the charge on each plate must be equal. Since the potential difference across each capacitor is $\frac{Q}{C}$, and the total potential difference across all of them in series is the sum of the potential differences across each individual capacitor, we can express an "equivalent capacitance" where

$$\frac{Q}{C_{eq}} = \frac{Q}{C_1} + \frac{Q}{C_2} + \frac{Q}{C_3} + \cdots \frac{Q}{C_n}$$

for the total potential difference across *n* capacitors in series. Because the charge *Q* is the same and therefore appears in every term, *Q* can be canceled out from each term, giving us the equivalent capacitance for capacitors in series as

$$\frac{1}{C_{eq}} = \frac{1}{C_1} + \frac{1}{C_2} + \frac{1}{C_3} + \cdots \frac{1}{C_n} \qquad (11\text{-}28)$$

Note that this looks very similar in form to the equation to find equivalent resistance for resistors in *parallel*, but this is an equation for capacitors in *series*. The question is naturally begged as to what the equivalent capacitance is for capacitors in *parallel*—capacitors with the same potential difference across them as each other—and in fact we find that it looks similar to the formula for resistors in series:

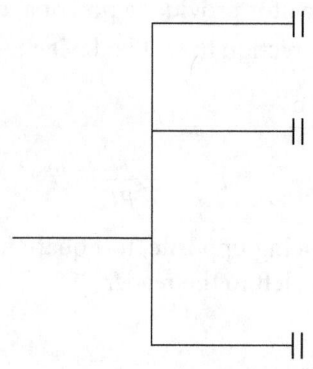

Here we can add together the charges on each plate of the capacitor to find the total charge accumulated, following the same principle as we used before of charge conservation (which in effect is the same principle as Kirchhoff's junction rule). Since the charge on each plate is $Q = C\Delta V$, we can write the total charge as $C_{eq}\Delta V$ for the "equivalent capacitance" C_{eq}, and

$$C_{eq}\Delta V = C_1\Delta V + C_2\Delta V + C_3\Delta V + \ldots + C_n\Delta V$$

As before, the potential difference ΔV is the same for all the capacitors, and therefore appears in every term, and can be canceled out:

$$C_{eq} = C_1 + C_2 + C_3 + \ldots + C_n \tag{11-29}$$

just like the formula for equivalent resistance in series.

To keep them straight from each other, we can write out the formulas for resistors and capacitors in series and in parallel in a table:

Table 11-1

	Series	Parallel
Resistors	$R_{eq} = R_1 + R_2 + R_3 + \cdots R_n$	$\dfrac{1}{R_{eq}} = \dfrac{1}{R_1} + \dfrac{1}{R_2} + \dfrac{1}{R_3} + \cdots \dfrac{1}{R_n}$
Capacitors	$\dfrac{1}{C_{eq}} = \dfrac{1}{C_1} + \dfrac{1}{C_2} + \dfrac{1}{C_3} + \cdots \dfrac{1}{C_n}$	$C_{eq} = C_1 + C_2 + C_3 + \cdots C_n$

Electrical Power

This chapter will be concluded with a brief word on electrical power. Although determining their resistance in the laboratory is easy, most commercially available resistors have their power consumption, rather than resistance, printed on the object. The best example of this would be a 60-W light bulb. A "watt" or "W" is the unit of power, and power is defined quite simply as

$$P \equiv \frac{W}{t} \tag{11-30}$$

where W is the work performed. Power is simply the rate of work performed, or (recalling that $W = -\Delta U$) it is the rate of energy consumed.

Electric potential energy is related to electric potential by $U = qV$, and the power is found by dividing both sides by t (or differentiating both sides by t, for the calculus-inclined). Doing so gives $P = \dfrac{U}{t} = \dfrac{qV}{t}$, but since $I = \dfrac{q}{t}$,

$$P = IV \tag{11-31}$$

Using Ohm's Law to make substitutions for I and V allows us to write P in terms of resistance, with either I or V depending on what we need:

$$P = I^2R \tag{11-32}$$

and

$$P = \frac{V^2}{R} \tag{11-33}$$

The reader should have no trouble performing either derivation.

Table 11-2 Table of Mass Densities and Resistivities for Some Common Materials at 20° C

Material	Mass Density	Resistivity
Carbon (C) (diamond)	3.515 g/cm^3	
Sodium (Na)	0.968 g/cm^3	47.7 nΩ m
Magnesium (Mg)	1.738 g/cm^3	43.9 nΩ m
Aluminum (Al)	2.70 g/cm^3	28.2 nΩ m
Potassium (K)	0.862 g/cm^3	72 nΩ m
Calcium (Ca)	1.55 g/cm^3	33.6 nΩ m
Titanium (Ti)	4.506 g/cm^3	420 nΩ m
Chromium (Cr)	7.19 g/cm^3	125 nΩ m
Manganese (Mn)	7.21 g/cm^3	1.44 μΩ·m
Iron (Fe)	7.874 g/cm^3	96.1 nΩ m
Cobalt (Co)	8.90 g/cm^3	62.4 nΩ m
Nickel (Ni)	8.908 g/cm^3	69.3 nΩ m
Copper (Cu)	8.96 g/cm^3	16.8 nΩ m
Zinc (Zn)	7.14 g/cm^3	59.0 nΩ m
Palladium (Pd)	12.023 g/cm^3	105.4 nΩ m
Cadmium (Cd)	8.65 g/cm^3	72.7 nΩ m
Cesium (Cs)	1.93 g/cm^3	205 nΩ m
Barium (Ba)	3.51 g/cm^3	332 nΩ m
Tungsten (W)	19.25 g/cm^3	52.8 nΩ m

HOMEWORK FOR CHAPTER 11

Name_____

A. Find the number density of atoms in the following substances:
   a. Iron

   b. Copper

c. Cobalt

d. Nickel

e. Aluminum

f. Cadmium

g. Tungsten

B. Suppose a current of 10 A were running through a circuit. If the circuit consisted of copper wires with cross-sectional radii 1.0 mm, what will the drift velocity of the charges be? Assume monovalent charge transfer.

C. An incandescent light bulb requires about 0.50 A of current to shine steadily (Giordano, *College Physics: Reasoning and Relationships* [Farmington Hills, MI: Cengage, 2009], p. 603). Tungsten filaments have a radius 0.023 mm, according to a fine intellectual journal available online (*http://en.wikipedia.org/wiki/Light_bulb#Filament*), and length 580 mm. Find the time it takes for electrons to move from one end of the filament to the other.

D. Resistance is a physical property of the wire or cable, related to resistivity ρ (a characteristic of the material) and to the length and cross-sectional area of the wire by the formula

$$R = \rho \frac{l}{A}$$

The resistivity of tungsten (ρ) is something you can look up. Find the resistance in the tungsten filament mentioned in problem C.

E. The typical light bulb dissipates 60 W of power. Power, as you may recall from Physics I, is defined as being the work done (in this case by a definitely nonconservative force) divided by time. It happens that for the case of electrical circuits, power is related to the potential difference induced and the current by the formula

$$P = IV$$

Most wall outlets provide a potential difference of 120 V. Find the current running through a 60-W light bulb.

F. Hurricane Broberg is about to hit the Georgia coastline, and you need to get out quickly before you get snowed into your home. (Hurricane Broberg is a Minnesotan hurricane. It snows instead of rains.) Fortunately, you are taking a physics class, and you are going to try to make your getaway by inventing a teleportation device. You know that the Humpty Dumpty Power Plant in Savannah has a power line going straight to San Diego, California, 3,800 km away, which should be a good safe distance. The wire is pure iron with a cross-sectional radius 0.5 mm. Your plan is the following: You're going to jump into the poorly secured coal-burning oven at the power plant and be incinerated. Just before jumping in, you're going to call a buddy in San Diego and have him turn his freezer on. You know that electricity works by moving electrons, so your dubious plan is to be un-incinerated when your electrons make it from the power plant to the freezer. Setting aside some of the technical difficulties that you may want to think through first (including the fact that most power grids are AC circuits!), how long will it take for the electrons from the power plant to reach your buddy's freezer in San Diego? *Hint:* To find the current, it may be useful to use Ohm's Law, $V = IR$. You can look up the resistivity of iron, and assume a potential difference of 120 V.

G. A 4 Ω, a 5 Ω, a 3.6 Ω, and an 8 Ω resistor are all in parallel. Find their equivalent resistance.

H. A 4 Ω, a 5 Ω, a 3.6 Ω, and an 8 Ω resistor are all in series. Find their equivalent resistance.

I. A 4 F, a 5 F, a 3.6 F, and an 8 F capacitor are all in parallel. Find their equivalent capacitance and find the potential difference across them if they are charged by a 0.50 A current for 3 minutes and 20 seconds.

J. A 4 F, a 5 F, a 3.6 F, and an 8 F capacitor are all in series. Find their equivalent capacitance, and find the charge on the positive plate of the 3.6 F capacitor if they are connected for a really long time to a 120 V wall outlet.

K. A lamp with a 60-W light bulb is plugged into a wall socket, which in North America delivers a current from a voltage difference of 120 V. Find (a) the resistance in the light bulb and (b) the current passing through it.

L. A 12-V battery is connected to a 3 Ω resistor, which is in parallel with a 7 Ω resistor. Both are in series with a 12 Ω resistor. What is the current that will pass through the 12 Ω resistor?

M. Find the currents in each of the resistors in the following circuit:

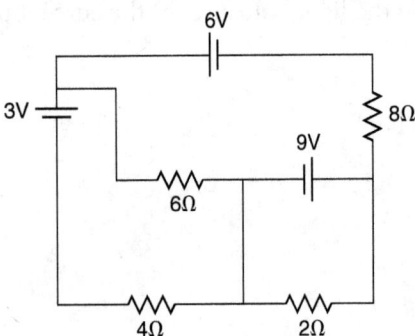

N. Find the currents in each of the resistors in the following circuit:

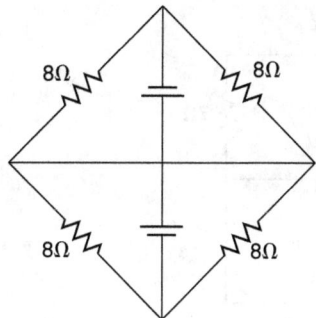

O. Find the currents in each of the resistors in the following circuit:

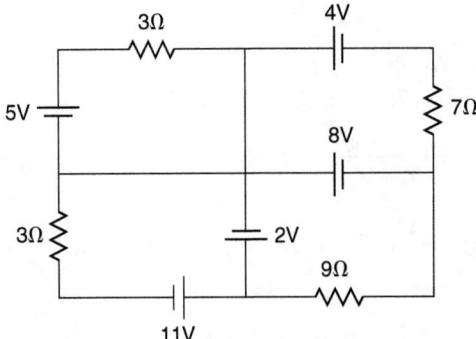

P. Find the currents in each of the resistors in the following circuit:

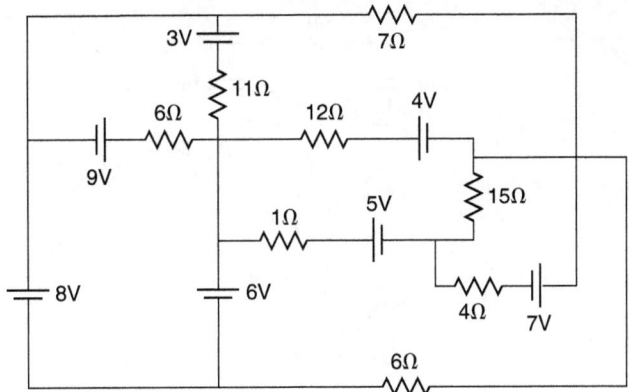

Q. Find the current through the resistor, and the charge on the positive plate of the capacitor, after 17 seconds:

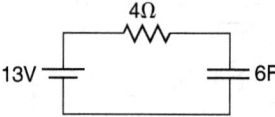

R. If one charged the capacitor in problem Q for a very long time, and then removed the battery while keeping the circuit closed, what will the current be 25 seconds after the battery is removed?

S. Find the current through the 7 Ω resistor, and the charge on the positive plate of the capacitor, after 30 seconds:

Magnetostatics 12

In Chapter 11, we mentioned that electric currents do not exist without the presence of magnetic fields. Loosely speaking, we might say that an electric current "causes" a magnetic field. More precisely speaking, all we can say is that the presence of a magnetic field encircling a current-carrying wire is the logical consequence of the presence of a current in that wire. And a careful thinker will remind us that that's not quite the same as saying that the current *causes* the magnetic field.

Magnetic fields have of course been known for millennia. Lodestone, a magnetized chunk of the mineral magnetite, is a permanently magnetized piece of rock found in nature that will align itself with the earth's magnetic field when suspended freely or left to float in a bowl of water. Although magnetized objects are electrically neutral, and *stationary* charged particles have no magnetic field associated with their charge, magnetization can resemble electric charge insofar as opposite poles (labeled "north" and "south") attract each other, while the same poles repel. There is a physical difference between the north and south poles, defined by the difference between positive and negative charges. If one draws magnetic field lines from the north to the south poles, defining them according to the direction that metal objects (which are temporarily magnetized) such as iron filings align themselves in the presence of a magnet, the direction of the magnetic field goes from the north pole to the south pole. And the direction of this magnetic field will determine the direction of the force exerted on a *positively* charged object moving in this field, as is explained in the section "Lorenz Force Law."

Like small pieces of lodestone, the Earth itself is a large magnet—its magnetic field actually being a consequence of free electrons in Earth's iron core rotating and thereby constituting a current, the same principle that we use to examine the magnetic fields associated with currents in electrical circuits. Like any magnetized object, Earth has both a north and a south magnetic pole—*every* object has both a north and a south pole. Although current research is suggesting that a very small number of "magnetic monopoles" may have been created in the early universe, and although the existence of "magnetic monopoles" will explain why electric charge in subatomic particles only comes in discrete quantities, none of them have ever or are expected to ever be observed in the laboratory.

Since opposite poles of a magnet attract, a piece of lodestone left to rotate freely will align itself along Earth's magnetic field lines. Because the lodestone's magnetic north pole points in a certain direction, that direction is known to us as "north" (although, since *opposite* poles attract, that's actually Earth's *south* magnetic pole).

While magnetic fields have been employed for navigation since before the birth of Christ, a scientific breakthrough in understanding their nature did not come until 1820, when Danish physicist Hans-Christian Ørsted, giving a lecture on batteries and electric currents, noticed that a compass needle in the vicinity of the wire was deflected when the current was turned on. A mathematical formulation of Ørsted's discovery came in 1827 when André-Marie Ampère, after whose honor the unit "Ampère" for current is named, published his *Mémoire sur la théorie mathématique des phénomènes électrodynamiques uniquement déduite de l'experience* (*Memoir on the Mathematical Theory of Electrodynamic Phenomena, Uniquely Deduced from Experience*). This treatise contained a presentation of what has come to be known as Ampère's Law. This law is consistent with, and can be derived from, a general expression for magnetic induction found in 1820 by Jean-Baptiste Biot and Felix Savart called the Biot-Savart Law, but this law is not easily presentable without some conceptually difficult vector calculus, and can be skipped over without impunity since it is Ampère's Law that is most useful for calculating magnetic induction.

Ampère's Law

Just as in Chapter 10 when we drew an imaginary surface enclosing all the charges which were creating an electric field, calling this a "Gaussian surface," now we draw an imaginary loop enclosing all the currents creating the magnetic field. If the loop is drawn in a nice symmetrical fashion so that the magnetic field is the same at every point on the loop, we'll be able to use Ampère's Law to easily calculate it. (It doesn't need to be the same everywhere on the loop—Ampère expressed his law as an integral equation which could describe non-symmetric problems. One is always welcome to do difficult and possibly impossible path integrals. However, despite the common misperception of many students, physics is actually supposed to make life *easier* for us, not harder—and why do integrals when you don't have to?)

In non-calculus-based terms, Ampère's Law says that

$$\mathbf{B} * \mathbf{L} = \mu_0 I_{enc} \tag{12-1}$$

where **B** is the "magnetic induction." "Magnetic induction" is a quantity that takes into consideration *both* the actual magnetic field strength itself—something whose value is independent of the matter inside of it, and to which we give the name **H**—and the magnetization of that matter inside of it, which in empty space will be 0. In empty space (which is all that we will concern ourself with in this book), magnetic field and magnetic induction are proportional and are often spoken of interchangeably, but they're actually not the same. They have different units, and are related by the equation

$$\mathbf{B} = \mu_0(\mathbf{H} + \mathbf{M}) \tag{12-2}$$

for the "magnetization" **M** of the matter inside the field, which is 0, for empty space. μ_0 is a constant, called the "magnetic constant" or "permeability of free space," and has a value $4\pi * 10^{-7}$ T m/A, where "T" stands for "Tesla" (the unit of **B**, named after Serbian-American physicist Nikolai Tesla) "m" is meters, and "A" is Ampères.

I_{enc}, not surprisingly, stands for the current enclosed by the "Ampèrean loop," and **L** is the *circumference* of that loop. Of course, more generally instead of taking the dot product between **B** and **L**, we're integrating **B** over that entire loop, which means performing an integral over the dot product **B** * d**L**, but the path integral is fortunately unnecessary if we ensure a constant **B** over the entire path. Likewise, I_{enc} is more generally speaking an integral of the current *density* passing through a given surface, but if the current comes from discrete wires that we can simply add together, that nasty surface integral (summing together an indefinite number of infinitesimal currents, rather than just a simple arithmetic sum of a few of them) is avoided as well.

As had been noticed by Ørsted after he started placing his compass at different points around the wire that was affecting its behavior, the magnetic field around a current-carrying wire will point in a circular shape, looping around the wire with a magnitude constant at any given radius. If one points the thumb of his right hand in the direction of the current—or sticks a pen in the cradle of his right hand pointing in the direction of the current—with the other fingers curling around the pen, the fingers will curl in the direction of the magnetic induction. (This will give the backward answer if done with the left hand, so it's called the "right-hand rule.") If the current comes toward us, out of a two-dimensional plane, it will be associated with a counter-clockwise magnetic induction. Likewise, if a current comes away from us into a two-dimensional plane, the magnetic induction will be clockwise. The direction of the current enclosed by a magnetic field is the same as the direction of the angular velocity and acceleration vectors discussed in Chapter 8, although current is *not* a vector (despite it having a direction, a vector must have a minimum of two dimensions, while a current I is bound to a one-dimensional wire, which itself will have a direction, but which could be bent in various directions without changing the current).

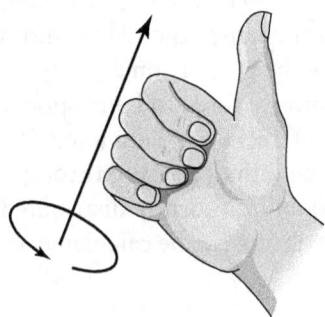

Since the magnetic induction is constant at fixed distances from the wire, and the locus of points in a fixed distance from the center is by definition a circle, the Ampèrean loop is going to be a circle, and the path length **L** is the circumference of a circle, $2\pi r$. Replacing L with $2\pi r$ and solving for B (which will only be a *scalar* quantity, since the right-hand side of the equation is a scalar, which is okay, because **B** and **L** are perpendicular to each other and are not added together, so all the vectors can be replaced with their full magnitudes), we get

$$B = \frac{\mu_0 I}{2\pi r} \qquad (12\text{-}3)$$

The magnetic field around a current-carrying wire is a logical consequence of the presence of a current through that wire, rather than vice versa. As we recall from Chapter 11, the charges in a DC circuit are usually caused by the electric force, not by the magnetic force. In looser language, the current "causes" the magnetic field, rather than the magnetic field causing the current. Physics itself of course can actually say nothing about what is "creating" or causing what, since a mathematical equation can only insist on consistency—"causality" has no mathematical definition. Many standard presentations of electrodynamics, such as David Griffiths's *Introduction to Electrodynamics*, define magnetic induction first, then proceed to the force on a charged particle associated with a magnetic field, and only introduce the concept of current later, within the larger context of magnetostatics.

Faraday's Law

The relationship between electric and magnetic fields actually has two aspects, both empirically observed. André-Marie Ampère's formulation of Ampère's Law came in 1827, showing that a steady current composed of moving charges will create a uniform magnetic field circling around the way; four years later in 1831, Michael Faraday showed that a changing magnetic field will in turn induce an electric potential difference in a conductor.

What Faraday actually found was that when he slid a bar magnet into and out of a loop of coil, a current began flowing through that coil—just as would happen if the coil had been cut and an electric potential difference induced by a battery between the ends. Although the presence of a magnet does not actually induce a physical potential difference between two points causing the charges to "fall" from one end to another, it does give a measurable kinetic energy to the charges moving around the loop; after all, *any* object with a velocity has kinetic energy. Faraday defined the energy given to each charge by the name "electromotive force"; this term has also been used in various places to refer to the energy given to each charge by a battery. We've been careful to avoid the term before—"electromotive force" is of course no force at all, since it doesn't have units of force—which would be Newtons—but rather it has units of Joules per Coulomb. And a Joule per Coulomb is of course a Volt, since $U = qV$. Force has units of (Volts * Coulomb)/m, since in general it also depends on the charge of the particle upon which the force is acting, which for a circuit will always be $1.6022 * 10^{-19}$ C, the magnitude of the charge on an electron, the only type of particle that carries charge in a circuit—and on the length traversed in the circuit. However, for any given circuit (in which the distance and charge on the charge carrier are always the same), the "electromotive force" and the actual force on the electrons in the wires will always be proportional.

Faraday found the electromotive force to be proportional to the rate of change in "magnetic flux." Recall that in Chapter 10, in our derivation of Gauss' Law, we had defined electric flux as the number of electric field lines passing through a given surface. When the electric field lines are perpendicular to and uniformly distributed through that surface (requiring a uniform electric field), the electric flux can be calculated as the value of the electric field at that point times the area, or

$$\Phi_E = EA \qquad (12\text{-}4)$$

The subscript "E" indicates that this is a definition of *electric* flux. Magnetic flux is defined the same way, provided the magnetic field is uniform (or that the magnetic field lines are uniformly distributed) through a given area to which they are perpendicular, the magnetic flux through an area is

$$\Phi_B = BA \qquad (12\text{-}5)$$

Of course, similar to electric flux, in general magnetic flux can be defined as the integral of the magnetic field over an area ($\int \mathbf{B} * d\mathbf{A}$), eliminating the requirement that the magnetic field lines be uniformly distributed or that they be perpendicular to the surface; however, computational examples where this surface integral needs to be performed can be easily ignored.

Faraday found that the rate of change of this entire quantity Φ_B is equal to the potential difference or "electromotive force" causing a current, allowing us to express Faraday's Law as the simple relation

$$\mathcal{E} = \frac{\Delta \Phi_B}{t} \qquad (12\text{-}6)$$

Because the change in Φ_B can result from a change in either B or A, the "electromotive force" can be seen whether we move a bar magnet in and out of the loop (changing B) or whether we rotate a wire loop inside of a magnetic field. Rotating a loop of wire changes the area presented for electric field lines to pass through, thereby changing the number of field lines passing through it—and changing the magnetic flux. Employing this fact by rotating a copper disk through a magnetic field, Faraday was able to create the first electromagnetic current generator. Modern-day power generators are nothing but more sophisticated applications of the same basic design, either using mechanical energy from steam to cause a magnetic rotor to spin inside of a wire loop (called a "stator"), or rotating coils of wire inside of a magnetic field in a design similar to Faraday's rotating copper disk.

We've chosen to use the symbol $\mathcal{E}$ for "electromotive force" in order to try to minimize confusion as much as possible. It's *not* a force—it has the same units as electric potential. But it's *not* an electrical potential either, since the force causing the electrons to move is not an electric force conveyed through an electric field whose origin is a charged particle as described by Coulomb's Law. Even though it only acts on something with an electric charge, which is also true of the electrostatic force, the force causing electrons to move through the circuit is a magnetic force, and the magnetic force acting on a charged particle cannot be found using Coulomb's Law.

Lorenz Force Law

Since Faraday's Law tells us everything about the apparent potential difference—and therefore the current—caused by a changing magnetic flux, one would expect that one could derive an expression for the actual magnetic force acting on the charge carriers from Faraday's Law itself. Indeed, this derivation can in fact be performed, and sixty years after Faraday's Law was presented to the scientific community, Oliver Heaviside derived an equation in 1889 that ended up being credited to the second physicist to derive it, Hendrik Lorentz. With a bit of vector calculus, Heaviside and Lorentz were able to show that the magnetic force on a charged particle moving at velocity v is

$$\mathbf{F_B} = q\mathbf{v} \times \mathbf{B} \qquad (12\text{-}7)$$

"B" is the same magnetic induction found using Ampère's Law, for which reason the subscript "B" is used to distinguish this force from the electric force.

Velocity and magnetic induction are both vectors, and this is a *cross product*, a way of multiplying two vectors together mentioned briefly before, but we won't always be able to take shortcuts as we could before (when calculating torque in Chapter 8, for example).

For any generic cross product (such as $\mathbf{A} \times \mathbf{B} = \mathbf{C}$), the magnitude of $\mathbf{C}$ is still going to be $AB \sin \theta$, where θ is the angle between the vectors $\mathbf{A}$ and $\mathbf{B}$. However, the *direction* of $\mathbf{C}$ will still need to be determined, and that can only be done in one step if $\mathbf{A}$ and $\mathbf{B}$ are perpendicular to each other, in which case we can use the "right-hand rule." If $\mathbf{A}$ and $\mathbf{B}$ are perpendicular to each other, then one can point one's thumb in the direction of $\mathbf{A}$ (and in our case, $\mathbf{A}$ will be $q\mathbf{v}$) and one's index finger in the direction of $\mathbf{B}$ (which in this case will be the magnetic induction). The third finger, pointing perpendicular to both $\mathbf{A}$ and $\mathbf{B}$, will indicate the direction of the cross product $\mathbf{A} \times \mathbf{B}$.

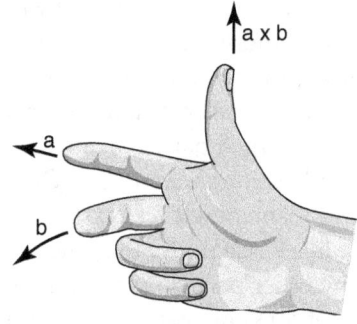

In general, $q\mathbf{v}$ is not going to be perpendicular to $\mathbf{B}$, and we can't use the right-hand rule to find the direction of $\mathbf{F}$. Instead, we have to find the components of $\mathbf{F}$. It can be proven that, for any cross product $\mathbf{A} \times \mathbf{B} = \mathbf{C}$, the components of $\mathbf{C}$ can be expanded by finding the (quasi-)determinant of the following 3×3 matrix:

$$\begin{vmatrix} \hat{x} & \hat{y} & \hat{z} \\ A_x & A_y & A_z \\ B_x & B_y & B_z \end{vmatrix}$$

where the cofactor expansion of the matrix (using Sarrus' Rule) gives us $\hat{x}(A_y B_z - A_z B_y) - \hat{y}(A_x B_z - A_z B_x) + \hat{z}(A_x B_y - A_y B_x)$ where $\hat{x}, \hat{y}$, and $\hat{z}$ are the unit vectors for their respective Cartesian coordinate. (A true determinant does not have vectors as elements in the array, which is why we were careful to call this a "quasi-determinant.")

To take an example, suppose a 6 nC charged particle is traveling at a velocity of $3.0 * 10^6$ m/s in the *x-y* plane along a 75° angle from the *x*-axis, through a magnetic field with induction **B** = 3.4**x** + 7.5**y** + 25**z**. If we were to use the Lorenz Force Law to find the force on that charged particle,

$$\mathbf{F_B} = q\mathbf{v} \times \mathbf{B}$$

would give us

$$\mathbf{F_B} = \det \begin{vmatrix} \hat{x} & \hat{y} & \hat{z} \\ qv_x & qv_y & qv_z \\ B_x & B_y & B_z \end{vmatrix}$$

$$= \det \begin{vmatrix} \hat{x} & \hat{y} & \hat{z} \\ (6*10^{-9})(3.0*10^6)\cos(75°) & (6*10^{-9})(3.0*10^6)\sin(75°) & 0 \\ 3.4 & 7.5 & 25 \end{vmatrix} =$$

$$\hat{x}\left((6*10^{-9})(3.0*10^6)\sin(75°)*25 - 7.5*0\right)$$

$$- \hat{y}\left((6*10^{-9})(3.0*10^6)\cos(75°)*25 - 3.4*0\right)$$

$$+ \hat{z}\left((6*10^{-9})(3.0*10^6)\cos(75°)*75 - (6*10^{-9})(3.0*10^6)\sin(75°)*3.4\right)$$

$$= 0.43\hat{x} + 0.12\hat{y} - 0.024\hat{z}.$$

One consequence of the magnetic force being a cross product is that it must be perpendicular to both **v** and **B**. The force of course points in the same direction as the acceleration—Newton's Second Law hasn't magically ceased to apply—and acceleration and velocity are never perpendicular for *linear* motion in a straight line where something speeds up or slows down, like our rocket problems and our free projectiles and our elevator problems and our equilibrium force diagrams. We *have* seen one case where velocity and acceleration were perpendicular, and that's in circular motion from Chapter 8. A magnetic force is in fact a great way to get particles moving around in a circle—it's how large particle accelerators or "cyclotrons," which need loops in order to accelerate particles across greater distances than would be feasible for linear tracks, keep charged particles moving in a perfect circle. The accelerator physics used to design such grand-scale accelerators as CERN (with some 5,000 magnets) and Fermilab is rooted, at its most fundamental level, on this basic fact.

A charged particle moving in a circle under the influence of a magnetic field has the same centripetal acceleration that any other particle moving in a circle does, and equating the formula for magnetic force with that for centripetal force, we see that

$$q\mathbf{v} \times \mathbf{B} = m\frac{v^2}{r}\hat{r} \qquad (12\text{-}8)$$

The unit vecto $\hat{r}$ is necessary so that the object on each side of the equation is in fact a vector.

The typical practical problem faced in a situation like this is determining what magnetic field is necessary to keep a charged particle moving in a circle of a given radius—or, alternatively, what the radius of a charged particle moving in a circle under a given magnetic field will be. For circular motion, **v** and **B** are perpendicular, so we can set the magnitudes of each side equal to each other and solve for B:

$$qvB = \frac{mv^2}{r}$$

$$B = \frac{mv}{qr} \tag{12-9}$$

If we chose to instead solve for r, telling us the radius that a particle moving in the presence of a magnetic field would take (a common example being for a particle spiraling around magnetic field lines in a plasma environment like the sun), we would get

$$r = \frac{mv}{qB} \tag{12-10}$$

which is the so-called "cyclotron equation."

We could also calculate the frequency of particles being accelerated in such a ring. The frequency of a particle looping around a circle is the number of times it passes through the loop per second, with units of sec^{-1}. It's a bit easier to think about if we find the time it takes for a particle to travel in a loop, or "period" of the oscillation—that's simple kinematics—and then take the reciprocal of that number to find the frequency.

Since velocity is defined as displacement divided by time, time must be equal to displacement divided by velocity, and we can solve the cyclotron equation for v to find a particle's velocity as a function of the particle's charge, mass, magnetic applied, and the radius of the loop:

$$v = \frac{qBr}{m} \tag{12-11}$$

Dividing the displacement $2\pi r$ (the circumference of the loop) by the velocity thus expressed, and following convention by using a capital rather than lowercase t for the period of the oscillation, a quantity that still is a measurement of time and has units of seconds,

$$T = \frac{2\pi m}{qB} \tag{12-12}$$

giving us an expression for frequency

$$f = \frac{1}{T} = \frac{qB}{2\pi m} \tag{12-13}$$

which, significantly, is *not* dependent on the radius of the cyclotron. This is because of Equation 12-11, where the velocity and radius are proportional to each other. The larger the radius of the loop, the less energy is lost through the work needed to constrain it into a circle, and the faster it will go, with the result that it completes the loop in the same amount of time as it had when the loop was smaller.

For example, let's see what the frequency of an electron being accelerated by a 1 T magnetic field would be. An electron has a charge $-1.6022 * 10^{-19}$ C and a mass of $9.109*10^{-31}$ kg. Since the frequency of the electron does not depend on the radius of the loop it is being accelerated through, we can just plug in the mass and charge of the electron to Equation 12-13 to see that $f = (1.6022*10^{-19})(1)/(2\pi * 9.109*10^{-31}) = 2.80 * 10^{10}$ s^{-1}.

While the Lorentz Force Law, giving the magnetic force on a charged point particle, is certainly of theoretical value in itself, and can be applied to situations such as the motion of subatomic particles through cyclotrons, or the motion of charged particles in the sun and other plasmas, the place where moving charged particles are most often to be found in day-to-day life is in a current. A current-carrying wire is in fact going to have lots of moving charged particles, and it would be most useful to be able to describe the magnetic force on the entire wire with one equation.

Fortunately, that's an easy thing to ask for. Nothing in the Lorentz Force Law specifies that q has to be a single charge, carried by a single "thing." Just as conservation of energy and momentum apply to *systems* of objects in Chapter 7, since there is no physical ground by which to distinguish a "thing" from its "parts", so q in the Lorentz Force Law can refer to the entire amount of charge which passes by a given point in a certain amount of time—or the entire amount of charge that passes by a given point in the time for that charge to move from one end of a wire to the other. This obviously is starting to sound reminiscent of current, since

$$I = \frac{q}{t} \tag{12-14}$$

allowing us to rewrite the charge upon which the Lorentz force is acting as

$$q = It \tag{12-15}$$

The time t here was said to be the time it takes for the electrons to move from one end of the wire to the other, where the total displacement they undergo is the length of the wire (which we can write as **L** instead of **Δx** for length, remembering that all we are talking about is the displacement of charges, which is a vector, whose *magnitude* is the length of the wire). The velocity **v** is related to the displacement **L** and the time t by the definition of velocity,

$$\mathbf{v} = \frac{\mathbf{L}}{t} \tag{12-16}$$

which fortuitously cancels out the "t" in the numerator, giving us

$$\mathbf{F_B} = It * \frac{\mathbf{L}}{t} \times \mathbf{B} = I\mathbf{L} \times \mathbf{B} \tag{12-17}$$

Once again, **L** is the displacement of charges in the wire, pointing in the direction of the current; the current itself is *not* a vector (it is the quotient of two scalars) because it is a one-dimensional quantity. A current does not change when a wire is bent; a current only changes at a junction.

Possible confusion should be eliminated here before it becomes a problem because we've seen the quantity **B** twice now in relation to currents. First we saw, using Ampère's Law, that current-carrying wires "create" or have associated with them magnetic fields looping around them, and now we are seeing the magnetic field exerting a force *on* a current-carrying wire. The magnetic field in the Lorentz Force Law is *not* the same field as the one "created" or associated with the magnetic field from Ampère's Law. It would violate Newton's Laws of motion for a particle or object to exert a force on itself, spontaneously pulling itself into motion with no reaction. You also can't draw an Ampèrean loop with radius 0 around a wire, dividing by 0 to get the magnetic induction by itself. Rather, **B** in Ampère's Law is the magnetic induction created by *another* wire (or by any other source). One current-carrying wire will exert a magnetic force on another one, and vice versa.

For example, consider two parallel current-carrying wires, such as shown here.

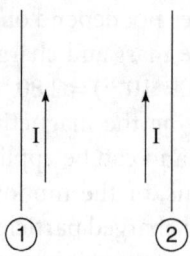

Wire 1 is going to create (or be associated with) a magnetic field $\mathbf{B}_1$ at the location where we've placed wire 2 exerting a force $\mathbf{F}_2$ on it, and wire 2 is going to create (or be associated with) a magnetic field $\mathbf{B}_2$ at the location where we've placed wire 1, exerting a force $\mathbf{F}_1$ on it.

Using the right-hand rule, we see that the magnetic induction $\mathbf{B}_1$ is coming down in the negative **z** direction at the location of the second wire, and that the magnetic induction $\mathbf{B}_2$ is coming up in the direction of the positive **z** direction at the location of the first wire. The force on each wire (labeled $\mathbf{F}_1$ for the force on the first wire and $\mathbf{F}_2$ for the force on the second wire) will be given by the Lorentz Force Law, where the force on each wire employs the magnetic induction created by the *other* wire:

$$\mathbf{F}_1 = I\mathbf{L}_1 \times \mathbf{B}_2 \qquad (12\text{-}18)$$

$$\mathbf{F}_2 = I\mathbf{L}_2 \times \mathbf{B}_1 \qquad (12\text{-}19)$$

The right-hand rule for the Lorenz Force Law has us point the thumb in the direction of **L** and the index finger in the direction of **B**, with the middle finger pointing in the direction of **F**. For $\mathbf{F}_1$, we see the thumb (current) pointing up the page in the $\hat{\mathbf{y}}$ direction and the index finger (magnetic induction from the other wire, $\mathbf{B}_2$) coming out of the page in the $\hat{\mathbf{z}}$ direction, leaving the middle finger pointing to the right toward the other wire. For $\mathbf{F}_2$, we see the thumb (current) pointing up the page in the $\hat{\mathbf{y}}$ direction and the index finger (magnetic induction from wire 1) coming into the page in the negative $\hat{\mathbf{z}}$ direction, leaving the middle finger pointing toward the *left*, which is also toward the other wire. Both forces push the wires toward each other, meaning that wires carrying parallel currents will attract each other.

Not surprisingly, currents with wires pointing the opposite direction—antiparallel currents—will repel each other. The same two equations, 12-18 and 12-19, will apply to the situation, but now the right-hand rule from Ampère's Law gives us $\mathbf{B}_2$ coming *into* the page in the $-\hat{\mathbf{z}}$ direction at the location of wire 1, while $\mathbf{B}_1$ as before still comes into the page in the $-\hat{\mathbf{z}}$ direction at the location of wire 2. Using the right-hand rule for the Lorentz Force Law, the force $\mathbf{F}_1$ has our thumb point *down* instead of up for the current, our index finger point into the page for the magnetic induction $\mathbf{B}_2$, and our middle finger consequently pointing to the *right*—away from the other wire. Likewise, the force $\mathbf{F}_2$ has our thumb point up for the current, our index finger point *into* the page for the magnetic induction $\mathbf{B}_1$, and our middle finger consequently pointing to the left—also away from the other wire. The two wires carrying antiparallel currents will repel each other.

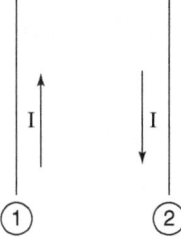

One final example can be given before the close of this chapter in order to illustrate the difference in scale between the electric and magnetic forces. In Chapter 10, we used Gauss' Law to find the *electric* field—and, by extension, the electric force—but a uniformly charged wire with some residual surface charge density λ. The currents introduced in Chapter 11 do *not* have a surface charge density. Electrons are being transferred from atom to atom, but the total number of electrons is the same as the total number of protons in the nuclei, and the wire as a whole is electrically neutral.[1]

---

[1] However, despite common misperceptions even repeated by professional physicists otherwise, there actually will be a net electric field outside the wire when a test charge is placed near the wire, because of the polarity in the wire induced by that charge, as discussed in the classic graduate text in electrodynamics, David Jackson's *Classical Electrodynamics*, and discussed specifically in the following paper: www.ifi.unicamp.br/~assis/Found-Phys-V29-p729-753(1999).pdf.

Since like charges repel each other, while parallel currents going the same direction will attract each other, we can take two *current-carrying* wires with a uniform line charge density λ and have the magnetic attracting force cancel out the electric repulsive force, keeping the wire in equilibrium. The uniformly charged wires we saw in Chapter 10 had static charges resting on them, and for simplicity's sake, let's assume a very slow drift velocity (almost always a true assumption), since the electric field from moving charges is not *quite* the same as the electrostatic fields we calculated in Chapters 9 and 10, and we don't want to have to worry about how the calculation changes.

In order to compare the magnitudes of the electric and magnetic forces, we could either be given the current passing through the wires and be asked to find the line charge density that one would need to place on each wire to keep them in equilibrium, or one could be given the value of the charge placed on the wire and be asked to find the current needed to overcome the electric repulsion. Let's be given the current—say, 0.5 A—and solve for the line charge density. We'll leave the length of and distance between the wires as an unknown variable, since as we'll see, the line charge density does not depend on either L or r—those values get canceled out.

We have two forces in equilibrium, so this is a Newton's Second Law problem. We draw a force diagram, and set up Newton's Second Law for the forces acting on the wire on the left:

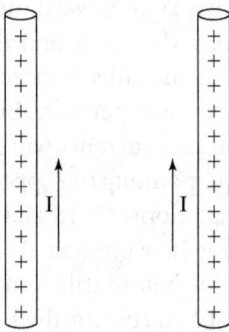

$$F_B - F_E = 0$$

F_B is given by the Lorentz Force Law, $I\mathbf{L} \times \mathbf{B}$, and F_E is going to be the total charge on the wire ($Q = \lambda L$) times the electric field created by the line charge, which as we recall from Chapter 10 is $E = \dfrac{\lambda}{2\pi r \varepsilon_0} = 0$. We're going to leave the length of each wire as L, and the distance between them as r:

$$I\mathbf{L} \times \mathbf{B} - \lambda \mathbf{L} = \left(\dfrac{\lambda}{2\pi r \varepsilon_0}\right) = 0 \qquad (12\text{-}20)$$

Both of these forces will be acting in the x direction, and $\mathbf{L}$ is perpendicular to $\mathbf{B}$, so the magnetic of the magnetic force is simply ILB. Using Ampère's Law to rewrite B as $\dfrac{\mu_0 I}{2\pi r}$ and multiplying it by IL,

$$\dfrac{\mu_0 I^2 L}{2\pi r} - \dfrac{\lambda^2 L}{2\pi r \varepsilon_0} = 0$$

L/r appears in both terms, so we can cancel them out. λ, the variable we are solving for, *does not depend* on either quantity. Keep in mind that we only have the current I squared because we specified that there would be the same current 0.5 A in both wires—one of those I's is the current from one wire, and the other is the current from the other.

$$\frac{\mu_0 I^2}{2\pi} - \frac{\lambda^2}{2\pi\varepsilon_0} = 0 \qquad (12\text{-}21)$$

While we're at it, we can cancel out the 2π's, and move the second term over to the right:

$$\mu_0 I^2 = \frac{\lambda^2}{\varepsilon_0} \qquad (12\text{-}22)$$

$$\lambda = \sqrt{\varepsilon_0 \mu_0 I^2} \qquad (12\text{-}23)$$

Now we can use a surprising fact discovered by Wilhelm Weber in 1856, and commonly misattributed to James Clerk Maxwell ever since, that the speed of light c is equal to $\frac{1}{\sqrt{\varepsilon_0 \mu_0}}$, a fact leading to the discovery that light waves are in fact nothing other than oscillating electric and magnetic fields. Using this fact, we can rewrite our line charge density as

$$\lambda = \frac{I}{c} \qquad (12\text{-}24)$$

The speed of light is 2.9979×10^8 m/s, *eight* orders of magnitude difference between the current in the wires causing the attractive force and the charge on the wires causing the repelling force. (Current and charge of course cannot be compared too strictly, since they have different units, but the only difference in unit is that one is divided by time and the other one isn't.)

HOMEWORK FOR CHAPTER 12

Name _____

1. A 5 kg object sits at rest in a magnetic field 0.01**y** T, where the unit vector **y** indicates the direction of the magnetic field. What is the magnitude of the magnetic force acting on the object?

2. One of the more important fundamental particles, the Brobergiton, has a mass $1 * 10^{-20}$ kg and charge $4.8 * 10^{-10}$ nC and is traveling at $2.7 * 10^7$ m/s along the y-axis in a magnetic field 0.035 $\hat{x}$ T. Find the magnitude and direction of the magnetic force exerted on the particle.

3. Suppose the Brobergiton were traveling at $2.7*10^7$ m/s along the y-axis in a magnetic field $0.035\ \hat{x} + 0.47\ \hat{y}$ T. Find the determinant of the cross product in the Lorenz Force Law to find the force exerted on the particle.

4. Suppose an iron wire cross-sectional radius 0.001 m is bent into a square loop with sides 0.3 m long, and is rotated with an angular velocity $\omega = 3\pi$ rad/s in a perpendicular magnetic field **B** = 0.5 T. Find the current induced through the wire. *Hint:* Ohm's Law holds true for "electromotive force" as well as for electric potential, since the electrons do not "know" what force is pushing them through the wire. You will have to find the resistance in the wire using the resistivity of iron. The rate at which the area of a rotating square loop with side d changes is $2d * v$, and you can refer back to Chapter 8 to find v from the angular velocity. Keep in mind that r is going to be the radius of the circle traced out by a *single point* on that loop as it rotates, rather than being the length of the side of the loop.

5. Suppose you have a summer internship at a particle lab, and you have an electron with mass $9.109 * 10^{-31}$ kg and charge $-1.6022 * 10^{-19}$ C, which your supervisor wants to be accelerated at $3 * 10^6$ m/s (1% of the speed of light) through a small cyclotron with radius 1 m. How strong a magnetic induction will you need to keep the electron moving in a perfectly circular path?

6. Suppose your small cyclotron with radius 1 m has an electron being accelerated by a 1 T magnetic field. What will the electron's frequency be?

7. Five wires with currents 3 A, 2 A, 1 A, 7 A, and 6 A are held closely together. Find the magnetic field 3 m away from the center using Ampere's Law.

8. Three wires with currents 7 A, –4 A, and 2 A are held closely together. Find the magnetic field 4 m away from the center using Ampere's Law.

9. Two wires each with current 3 A are held parallel to each other a distance 0.02 m apart. Find the magnetic force on one of the wires. Is it attractive or repulsive?

10. Two 5 m long wires each with current 3 A are held *anti*-parallel (with the currents going the opposite direction) a distance 0.05 m apart. Find the magnetic force on one of the wires. Is it attractive or repulsive?

11. Two 5 m long wires each with current 3 A are held perpendicular to each other a distance 0.04 m apart. Find the magnetic force on one of the wires.

12. Two 5 m long wires each with current 3 A are held, separated by a 40° angle apart in the *x-y* plane, with a vertical distance of 0.01 m between them along the *z*-axis. Find the magnitude and direction of the magnetic force on one of the wires.

13. One 5 m long wire with current 3 A is held parallel to another wire of the same length with current 5 A at a distance 0.03 m. Find the magnetic force on **each** wire.

14. Two charged conducting wires each have a current 7 A, and are placed parallel to each other. Find the line charge density that needs to be placed on each wire so that the electric repulsion between them balances the magnetostatic force between them so that they are in equilibrium.